CLASS EXPERIMENTS IN PLANT PHYSIOLOGY

CLASS EXPERIMENTS IN PLANT PHYSIOLOGY

HANS MEIDNER

*Department of Biological Science,
University of Stirling*

London
GEORGE ALLEN & UNWIN
Boston   Sydney

George Allen & Unwin (Publishers) Ltd,
40 Museum Street, London WC1A 1LU, UK

George Allen & Unwin (Publishers) Ltd,
Park Lane, Hemel Hempstead, Herts HP2 4TE, UK

Allen & Unwin Inc.,
9 Winchester Terrace, Winchester, Mass. 01890, USA

George Allen & Unwin Australia Pty Ltd,
8 Napier Street, North Sydney, NSW 2060, Australia

First published in 1984

British Library Cataloguing in Publication Data

Meidner, Hans
    Class experiments in plant physiology
1. Plant physiology—Laboratory manuals
I. Title
581.1'028   QK714.5
ISBN 0-04-581015-X
ISBN 0-04-581016-8 Pbk

Library of Congress Cataloging in Publication Data

Meidner, Hans.
    Class experiments in plant physiology.
Includes index.
Bibliography: p.
1. Plant physiology—Experiments. I. Title.
QK714.4.M45   1984   581.1'07'8   83–25693
ISBN 0-04-581015-X
ISBN 0-04-581016-8 (pbk.)

Set in 9 on 11 point Melior by
D. P. Media Limited, Hitchin, Hertfordshire
and printed in Great Britain by Mackays of Chatham

Preface

Class experiments in plant physiology has been written to provide a source book of technical information needed by teachers aiming to carry out class experiments successfully. For such an enterprise a single author would have serious limitations and it was therefore decided to invite the co-operation of colleagues in the United Kingdom so that it would become possible to compile a collection of tested class experiments. In order to present the contents in a uniform style these have been rewritten and edited by the author with the consent of the contributors, using the information they supplied.

The experimental topics range from physicochemical processes to enzyme- and hormone-mediated biochemical changes at the cellular and whole plant levels, covering most subdivisions of plant physiology. A notable omission is tissue culture which proved to be too large and specialised a subject – about which, however, a specialist book is available (Dodds 1983).

The value of experimentation is not fully realised by merely carrying out the experimental steps. Keeping proper practical class protocol and writing experimental reports so that others can repeat the experiment are equally important. Thus, advice on the preparation of experimental reports forms a short section in Part A of this book.

The author realises that most teachers of plant physiology excel in their mastery of the theoretical basis of plant processes. Therefore no attempt has been made to give theoretical background for the experiments, except for introductory remarks. However, the mastery of conducting class experiments in plant physiology is not equally well developed. A compilation of experimental accounts emphasising technical and experimental detail and underlining possible faults will improve the teaching of plant physiology as an experimental science.

REFERENCE

Dodds, J. D. 1983. *Experiments in plant tissue culture*. Cambridge: Cambridge University Press.

Acknowledgements

Sincere thanks are due to all contributors, whose excellent co-operation and prompt response made this compilation possible. Thanks are due, also, to my wife who has worked hard to make the passages of continuous prose render clearly what they set out to say; and I wish to acknowledge the outstandingly capable editing by Celia Caspell and the publishers of a very complex typescript.

In addition, the following organisations and individuals are thanked for permission to reproduce illustrative material:

N. Lewis (Table B2.5); Table B4.10 reproduced from *Plant physiology*, 2nd edn (R. G. S. Bidwell) by permission of the author and Macmillan Publishing, © 1979 R. G. S. Bidwell; M. W. Nabors (Figure B5.5b); Figure B11.11 reproduced from *What's New in Plant Physiology* **12**, 9–12 by kind permission of G. Fritz and T. J. Mulkey.

Contents

Part A   Introduction

About this book

Aims

The aim of compiling a source book of class experiments in plant physiology is to provide descriptions of well tried procedures from which instructors can extract the necessary information for specific experiments. The book is not intended to provide a 'set' course of class experiments. Uniformity of teaching courses is not thought to be desirable but rather variability and individually designed programmes. Hence this is a *source book* and not a schedule.

Where I could identify the originator of an experiment, I have acknowledged this; references to papers are quoted to indicate where methods have been developed, usually for research work, and where further information can be found. However, many procedures have become common property, originators are unknown or basic experimental arrangements were developed by different people in different places. Acknowledgement is therefore due to many anonymous plant physiologists.

There are comparatively simple exercises and very complicated ones; some more suitable for small groups, others for larger classes. All kinds have their place in teaching an experimental approach to plant physiology. They can form the basis of class discussions in which students combine their theoretical knowledge with their laboratory experience. Here and there, well known 'classic' experiments have been mentioned but not dealt with in detail as they can be found elsewhere. However, the view is not shared that only 'new' experiments are of value; some 'old' experiments are most instructive and have the advantage of being within the reach of laboratories not equipped with complex instruments. Where some well known experiments have been described in detail, hints are given for improvements in method or apparatus, which will make such experiments more successful than has often been the case.

Method of presentation

This book has been written with the aim of providing instructors with tested experimental procedures from which they can select appropriate experiments fitting their teaching programmes.

Part A contains very basic topics, namely the system of units used in this book, descriptions of common laboratory equipment, some ground rules for the preparation of laboratory reports and diagrams of apparatus, as well as procedures for collecting, growing and preparing plant material for physiological experiments. Thus Part A may be thought to be suitable for student use, but it has been included here to remind instructors that matters which have become 'second nature' to them need teaching thoroughly – though not in a doctrinaire manner.

Part B contains accounts of experiments, grouped under 11 headings. The style in which the accounts are presented is somewhat concentrated, to fit the subject matter into the available space but also intending that instructors should prepare their instructions or experimental schedules for their classes in their own style. However, the accounts are sufficiently detailed and provided with practical hints to be suitable not only for experienced teachers but also for those beginning their teaching career.

Several contributors who remarked on the fact that the accounts of some of the experiments might appear simplistic or incomplete supported their contributions with suggested questions to be put to the students for discussion both before and after results are obtained. This may well be the best method to stimulate students to think about the scheduled work, but it is left to instructors to devise such questions and they have not been included. As often as possible, the theoretical basis and context of the experiments has been briefly indicated, but it has been assumed that instructors are best suited to integrate the laboratory work with their lecture courses. Where appropriate and within the space available, relevant anatomical studies have been included so that the unity of structure and function can be emphasised.

Part C is in the nature of a technical directory. It contains formulae of those reagents and other preparations which are common to many experiments, addresses of suppliers of specialist materials referred to in the experimental accounts, and addresses of those colleagues who have contributed to this volume.

How to use this book

The form in which users will extract information from the book and give it to a class must vary with the nature of the course and the level at which it is given. It must also vary with the time of year and the climatic conditions in which an experimenter works. In certain cases experiments require some weeks for completion, but most are designed for class work of 3 h duration; therefore, it may appear that methods have been simplified and work is carried out, for instance, with crude instead of purified extracts (e.g. Expts 11.4 & 11.5). This should be explained to the classes. Many of the experiments described can be combined into sets in a particular subject area; others can be developed into small projects.

Classifying experiments under specific headings tends to misrepresent the subject matter and to introduce a risk that users of the book may fail to find some relevant part of the contents. Topics in plant physiology are interrelated, and whether a feature of germination, for instance, belongs to 'Growth', 'Respiration', 'Enzymes' or to 'Translocation' is almost an arbitrary decision. To list the same experiment under several of the most relevant headings would be wasteful of space and, in the end, confusing. Therefore, special attention should be paid to the cross references quoted at the head of most experiments, in the text and in the Index.

Every effort has been made to give an accurate account of the quantities of materials, reagents, experimental conditions and direction of change to be expected. All units used in this book are SI units; their relevant non-SI equivalents are shown in Table A1 and are often quoted in parenthesis in the individual accounts. It has been assumed that instructors are familiar with the working and calibration of complex equipment. Different makes of apparatus require different instructions and it would be impracticable to attempt to give instructions for using such apparatus as:

atomic absorption spectrophotometers	infra-red gas analysers
chromatographic equipment	oxygen electrodes
electrophoretic apparatus	porometers
flame photometers	psychrometers
gas chromatographs	scintillation counters
Geiger tubes	spectrophotometers
Gilson respirometers	Warburg apparatus

The marginal notes which are added to the descriptions of many experiments contain points of procedure learnt from experience. They draw attention to pitfalls and snags, and offer tips which should help to make the exercises successful.

My major concern is that anyone attempting to carry out with his class an experiment described in this text should succeed. This is difficult to ensure because experiments in plant physiology use variable materials and therefore their outcome is not entirely predictable. Environmental and laboratory conditions also vary not only with season but from day to day, and technicians and instructors are liable to err in the preparation of plant material and reagents, which can also deteriorate in storage (e.g. Expts 1.9, 1.10 & 2.3). These uncertainties are no reason, however, for class experiments to be inconclusive. There is one indispensable rule that alone can ensure successful experimentation:

Before a class begins, the instructor must carry out the essential steps of the experiment. This can be done in the morning before an afternoon class or early in the morning before a class starts and, if that is not possible, on the day before a class. Such checks do not take very long for an experienced instructor and he will discover any shortcomings, if these have occurred.

I have tested the essential steps in most of the accounts. If procedures are closely followed, these experiments should be successful, provided also that the plant material is in sound condition. For this reason I stress again the general rule stated above that few, if any, class experiments will succeed unless the essential features are tested immediately prior to laboratory classes.

SI units

The Système International d'Unités has been agreed upon with the aim of providing a system of units free from ambiguities and possibilities of misinterpretation. With this aim in mind SI units are used; however, it is not the aim here to present the system in its entirety, but only those parts of it which are used in this book.

Table A1 is a reference list of selected quantities, their SI units, symbols and dimensions, together with equivalent non-SI units and conversion factors where appropriate. Because traditional usage cannot be changed suddenly, SI units in the text have frequently been supplemented with conventional non-SI units in parenthesis.

Table A1 SI units and prefixes used in this book.

Prefixes common to all units		
10^6	mega	M
10^3	kilo	k
10^{-1}	deci	d
10^{-2}	centi	c
10^{-3}	milli	m
10^{-6}	micro	μ
10^{-9}	nano	n

Physical quantity	Name of unit	Symbol for unit	Definition or dimension of unit	Conversion	Non-SI unit
time (basic)	second	s	the duration of 9 192 631 770 periods of the radiation from a caesium-133 atom on transition between specified levels		
	minute	min			
	hour	h			
length (basic)	metre (meter)	m	1 650 763.73 wavelengths of radiation of a krypton-86 atom on transition between specified levels	2.54×10^{-2} m = 1 in	inch (in)
	centimetre	cm		2.54 cm = 1 in	
	millimetre	mm		25.4 mm = 1 in	
	micron	μm			
	nanometre	nm		10^{-1} nm = 1 Å	Ångstrom (Å)

Physical quantity	Name of unit	Symbol for unit	Definition or dimension of unit	Conversion	Non-SI unit
area	square metre	m²			
	square centimetre	cm²		6.4516 cm² = 1 sq. in	square inch (sq. in)
	square millimetre	mm²			
volume	cubic metre	m³		} numerically equal	{ kilolitre (kl)
	cubic decimetre	dm³			litre (!)
	cubic centimetre	cm³			millilitre (ml)
	cubic millimetre	mm³			microlitre (µl)
mass (basic)	kilogramme	kg	equal to the mass of an international prototype	0.4536 kg = 1 lb	pound avoirdupois (lb)
	gramme	g			
	milligramme	mg			
mass concentration			$kg\ m^{-3}$ $g\ dm^{-3}$ } $mg\ cm^{-3}$ $\mu g\ mm^{-3}$ }	numerically equal	{ $kg\ kl^{-1}$ $g\ l^{-1}$ $mg\ ml^{-1}$ $\mu g\ \mu l^{-1}$
special case for water vapour in air	water vapour density		$g\ m^{-3}$ } $\mu g\ cm^{-3}$ }	not readily convertible	{ atmospheric moisture content; % relative humidity; saturation deficit
special case for CO_2 in air	density of CO_2 in air		$mg\ m^{-3}$ $cm^3\ m^{-3}$	$2\ mg\ m^{-3} = 1$ p.p.m.	parts per million (p.p.m.)
amount of substance (basic)	mole	mol	amount of substance containing as many elementary units as 12 g carbon-12 units can be atoms, ions, molecules or photons (cf. light)		
	millimole	mmol	10^{-3} mol		
	micromole	µmol	10^{-6} mol		
	nanomole	nmol	10^{-9} mol		
amount of substance concentration molal	10^3 mole dissolved in 1 m³ water *or* 1 mole dissolved in 1 dm³ water		$10^3\ mol\ m^{-3}$ H_2O } $1\ mol\ dm^{-3}$ H_2O }	numerically equal	{ 10^3 mole dissolved in 1 kl water ($10^3\ mol\ kl^{-1}$ H_2O) 1 mole dissolved in 1 litre water ($1\ mol\ l^{-1}$ H_2O)

Physical quantity	Name of unit	Symbol for unit	Definition or dimension of unit	Conversion	Non-SI unit
molar: (a) based on conc. per cubic metre	10^3 mole per cubic metre of solution		10^3 mol m^{-3}		1.0 M 1 mol l^{-1}
	10^2 mole per cubic metre of solution		10^2 mol m^{-3}		0.1 M 10^{-1} mol l^{-1}
Molar solutions are of great value for class experiments	10 mole per cubic metre of solution		10 mol m^{-3}		10 mM 10^{-2} mol l^{-1}
because their dilutions result in	1 mole per cubic metre of solution		1 mol m^{-3}	numerically equal	1.0 mM 10^{-3} mol l^{-1}
known molar concentrations with known	10^{-1} mole per cubic metre of solution		10^{-1} mol m^{-3}		100 μM 10^{-4} mol l^{-1}
fractions of the solute potential of a 10^{-3} mol m^{-3} (1.0 M) solution.	10^{-3} mole per cubic metre of solution		10^{-3} mol m^{-3}		1 μM 10^{-6} mol l^{-1}
molar: (b) based on conc. per cubic decimetre	1 mole per cubic decimetre		1 mol dm^{-3}		1.0 M 1 mol l^{-1}
	10^{-1} mole per cubic decimetre		10^{-1} mol dm^{-3}		0.1 M 10^{-1} mol l^{-1}
	10^{-2} mole per cubic decimetre		10^{-2} mol dm^{-3}		10 mM 10^{-2} mol l^{-1}
	10^{-3} mole per cubic decimetre		10^{-3} mol dm^{-3}	numerically equal	1.0 mM 10^{-3} mol l^{-1}
	10^{-4} mole per cubic decimetre		10^{-4} mol dm^{-3}		100 μM 10^{-4} mol l^{-1}
	10^{-6} mole per cubic decimetre		10^{-6} mol dm^{-3}		1.0 μM 10^{-6} mol l^{-1}
energy	joule	J	kg m^2 s^{-2}	0.241 J = 1 cal 10^{-7} J = 1 erg	calorie (cal) erg

Physical quantity	Name of unit	Symbol for unit	Definition or dimension of unit	Conversion	Non-SI unit
force	newton	N	kg m s^{-2} (= 1 J m^{-1})	10^{-5} N = 1 dyn	dyne (dyn)
pressure (force per unit area)	pascal	Pa	N m^{-2}	10^5 N m^{-2} = 1 bar	bar
	megapascal (this unit is used for water, solute and turgor potential)	MPa	10^6 N m^{-2}	0.1 MPa = 1 bar	bar
				0.1013255 MPa = 1 atm	atmosphere (atm)
	kilopascal (unit used for atmospheric vapour pressure)	kPa	10^3 N m^{-2}	0.1 kPa = 1 mbar	millibar (mbar)
				0.133 kPa = 1 mmHg	millimetre mercury (mmHg)
power	watt	W	kg m^2 s^{-3} (= J s^{-1})		
light supply total irradiance (*must be measured as such*)	power per square metre		Wm^{-2}	*not* readily convertible	foot-candle (ft-candle) lux
				70 W m^{-2} = 1 cal m^{-2} min^{-1}	calorie per square metre per minute (cal m^{-2} min^{-1})
photosynthetic active radiation (PAR; *must be measured as such*)	photon-flux density		mol m^{-2} s^{-1}	numerically equal	einstein per square metre per second (E m^{-2} s^{-1})
			µmol m^{-2} s^{-1}		microeinstein per square metre per second (µE m^{-2} s^{-1})

Basic equipment

How to describe it

Descriptions of assembling apparatus can be lengthy without being complete and it is therefore good practice to supplement these with well labelled and captioned diagrams. As a rule, longitudinal sections of apparatus are adequately informative and only on rare occasions are perspective diagrams needed to clarify a three-dimensional aspect (e.g. Fig. B7.3). Likewise, there is rarely any need for shading or colouring; where this occurs in this book, it is for reasons of design.

Some rules for the correct representation of standard pieces of apparatus are as follows:

(a) In longitudinal sections rods are drawn with the ends closed and tubes with the ends open. If the thickness of the wall of the tube is important, double lines closed at the end should be used; this is essential for capillary tubes as shown in Figure A1.

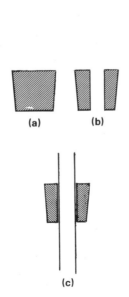

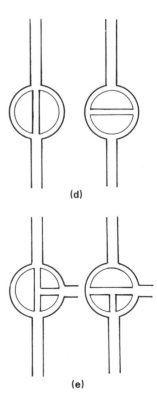

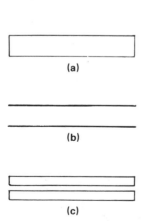

Figure A1 Longitudinal sections of (a) rod, (b) tube, and (c) capillary tube.

Figure A2 Longitudinal sections of (a) blind bung, (b) drilled bung, (c) tube inserted through drilled bung, (d) open and closed two-way stopcocks and (e) open and closed three-way taps.

(b) Blind and drilled rubber bungs or corks and stopcocks are shown in longitudinal section in Figure A2. For diagrams of complete experimental assemblies, those accompanying the experimental write-ups in Part B, Sections 1–11, can be taken as models, except that in a printed text differential shading is customary.

How to use it

atmometers (porous pots)

The essential part of the atmometer and allied instruments is the porous pot (see Fig. A3a). If used for gas diffusion (Expt 1.2), it must be absolutely dry; if accidentally wetted, a thorough washing and lengthy drying may restore it, but only tests will tell.

If the porous pot is used as an evaporator or liquid diffuser, it must be boiled in distilled water for at least 10 min, allowed to cool in the same water and filled with that water to the brim. Then, while the pot remains immersed in the water, the rubber bung with the tube already fitted through it is pushed firmly into the open end of the pot. These operations could be done in a filled sink. Prepared in this way, the porous pot functions perfectly, but never should an air bubble be allowed to enter it via the end of the tube, which must be kept dipping into water. If mercury

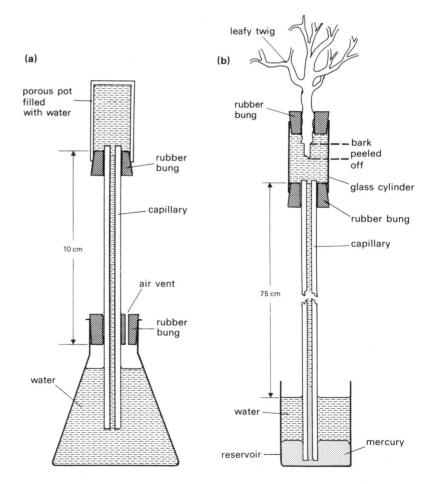

Figure A3 (a) Longitudinal section of assembly for an atmometer. (b) Longitudinal section of assembly for demonstrating 'transpiration pull'.

is to be used (cf. Expt 9.1), the porous pot, or the twig assembly, must first be set up with water and the mercury poured into the reservoir (see Fig. A3b) once water is moving up the tube.

If air has entered the pot, it can be reconditioned by boiling once more.

blades (scalpels) Good sections, even thick ones, can only be obtained with sharp blades. Only new blades are sufficiently sharp and, although it appears wasteful, for sectioning or making cuts into leaf tissue for epidermal peels, only a new blade will give good results. Scalpels, although very sharp, are meant for dissecting animal tissue in which cells have no cellulose walls; they are not made for botanical sectioning. A scalpel cuts by pressure and 'ripping'; sectioning is achieved by pressure combined with a skilled sliding movement of the blade across the material. For all the work mentioned in this book, new blades should be used, of either the single or double edged type.

brushes When handling material with forceps or needles, there is a danger of crushing or otherwise damaging delicate structures (Expt 5.5). The use of camel-hair brushes is therefore recommended, especially for some staining techniques (Expt 2.4 anatomy) and for the transfer of some material from one medium or vessel to another.

bungs, rubber To drill holes into rubber bungs, both the bung and the cork-borer must be kept wet all the time; best results are obtained if water plus a trace of surfactant is used.

burettes The use of burettes requires their filling and rinsing in class laboratories; this should be done via funnels. For some experiments (e.g. Expt 2.4) the scale has to be read in reverse – this can cause confusion.

camel-hair brushes see **brushes**

dark-field microscope Commercial dark-field condensers, if available, can often be used to advantage (Expt 1.1). The main point about their use is that they must be properly centred in the substage and their contact with the microscope slide from below must be accomplished by a drop of immersion oil, or glycerol in some cases, on their surface. For oil-immersion objectives a drop of immersion oil must, of course, also be placed between the cover-slip and the objective.

However, a dark-field effect can be produced with any condenser if it has a suitable holder above the iris diaphragm, by placing a disc as shown in Figure A4 and adjusting the iris diaphragm so that only a narrow ring of light appears round the central black disc. If there is no holder for such

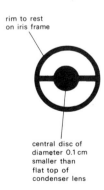

rim to rest
on iris frame

central disc of
diameter 0.1 cm
smaller than
flat top of
condenser lens

Figure A4 Plan of black-out spot producing a dark-field effect if placed underneath an ordinary condenser.

a disc above the iris diaphragm, a disc 0.1 cm smaller in diameter than the flat top of the condenser can be placed centrally on it. There is no need for oil when the latter arrangement is used.

eyepiece graticules   see **graticules**

filters   see **light filters**

forceps   The use of forceps for handling plant material is essential in many experiments – fingers should not be used (cf. Expt 10.11). However, very delicate structures should be handled with **brushes**.

glass tubing   Many experimental assemblies require air- and watertight joints between glass and rubber or plastic tubing. The parts to be joined should be a good fit in the first place. The joining is most readily achieved by assembling the pieces when wet or, indeed, with the pieces submerged in a sink full of water.

   If the correct size of tubing is not available, it is often just as effective to fit the tubing *into* the bore of the glass tube. If the wall of a rubber tube is turned over on itself by 0.3–0.4 cm (see Fig. A5), this will marginally reduce its bore so that a tighter fit is achieved. This cannot be done with plastic tubing.

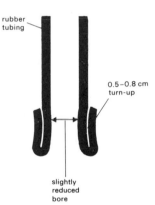

rubber tubing

0.5–0.8 cm turn-up

slightly reduced bore

Figure A5 Longitudinal section of rubber tubing turned over at its end to reduce its internal bore.

   The use of tubes of several different diameters, one fitted into the bore of another, effectively produces a reducing joint, but at all times it is essential to lubricate with water. On no account should Vaseline be used; although it smoothes the assembly, it also allows it to come apart easily and it is *not* a seal for water.

graticules and stage micrometers   Microscopes are often used as precision measuring instruments. An eyepiece graticule, i.e. a microscopic measuring scale, can be used with ease at 100×, 400× or 1000× magnification. Some counts and measurements of distances at 100× magnification are more manageable if the microscope field is imagined as divided in half or into a quarter of its area. The calibration of the eyepiece graticule with a stage micrometer may not be practicable in a large class as it requires many stage micrometers. In this case students should be advised of the measurement represented by the distance between two adjacent lines of the graticule, i.e. the smallest unit of the eyepiece graticule at the particular magnification used. Although objectives are usually uniform in class microscopes, the eyepieces may differ, so that for 10× a different calibration is required from that for 8× eyepieces. Eyepiece graticules should be available in reasonable numbers: at least one for each pair of students. They should also be used to calculate the areas of low and high power microscope fields which are required for cell counts (Expts 5.6b, 8.5 & 11.13 anatomy).

To calibrate the eyepiece graticule, a stage micrometer is needed. The scale on the stage micrometer is an accurately drawn 100 μm scale with 50 or 100 divisions, depending on quality and price. Every 10th or 20th division is numbered. The eyepiece graticule is usually a 1 cm scale with 100 divisions, of which every 10th is numbered. Calibration of the eyepiece graticule is accomplished by superimposing its scale at low and high magnification on the stage micrometer scale. The two will not match perfectly. For instance, under 400× magnification 27 divisions of the eyepiece graticule may correspond to 100 μm on the stage micrometer scale: therefore, one eyepiece graticule division equals 3.7 μm. At 100× magnification seven eyepiece divisions may correspond to 100 μm, i.e. each eyepiece division equals 14.3 μm and not $4 \times 3.7 = 14.8$ μm. The magnifications indicated for optical microscopes are not perfectly accurate: 400× magnification may only be 390×. Therefore, calibration should be carried out at each magnification separately. Calculating one from the other is liable to introduce errors.

light filters
Special filter arrangements are mentioned in Experiments 8.6, 10.12 and 11.13. However, Cinemoid colour filters used for theatre stage illumination are suitable for experimental use to produce broad-band blue, red and far-red light, the last being produced by combining the blue, orange and red filters. The colours usually required are: Deep Orange, no. 58; Ruby, no. 14; Primary Deep Blue, no. 20; and Primary Green, no. 39. The filters can be cut with scissors from sheets supplied by Rank Strand Electric Co. (see Part C2). Information on the transmission characteristics of the filters will be given by the suppliers.

light sources
For many purposes laboratory bench lamps are quite adequate. Mirror-backed reflector light bulbs are often useful but, for intense photon-flux densities, slide-projector bulbs are recommended (Expts 7.2 & 8.6).

porous pot
see **atmometer**

potometers
These instruments measure water inflow rates, *not* transpiration rates; they can be home-made and require only a horizontal capillary tube in which the rate of movement of an air bubble can be measured (Expts 3.6, 3.7, 5.2 & 5.3).

rubber bungs
see **bungs**

scalpels
see **blades**

stage micrometer
see **graticules**

syringes
Disposable syringes are reasonably accurate measuring instruments which can be incorporated into many assemblies of apparatus into which known quantities of reagents must be introduced without the need to dismantle the system (Expt 2.4). Syringes have the advantage also that once they are filled with a substance it remains sealed from the atmosphere (Expts 2.6, 2.8 & 8.4). Another valuable feature of syringes is the fact that syringe needles can be introduced via plastic or rubber parts while the apparatus remains gas-tight and once fitted they can remain sealed in (Expts 5.2 & 5.3).

thermocouples

For the measurement of temperature of tissues and of pieces of apparatus, thermocouples are a reliable means. Thermistors, where available, are also good, but they are more complex. The most convenient thermocouple is made from copper and constantan wires, which are readily available in many thicknesses. Whereas the copper takes the solder easily, constantan is on occasion difficult to solder. A little flux helps and the easiest way to solder the wires together is to twist them around each other and afterwards file the soldered junction flat and into a sharp point that can be inserted into any tissue. The temperature that is measured is the temperature at the junction of the two wires as shown in Figure A6, which also illustrates the circuit. Constantan usually forms the shorter loop provided it is long enough to prevent temperature conduction between the two junctions. In practice, the constantan loop must be long enough to allow for the placing of the reference junction in its constant temperature container and the measuring junction in the plant material.

The output from the thermocouple is measured with a microvoltmeter. Calibration is accomplished by keeping the reference junction at a constant temperature, for instance 0 °C, provided by ice-water monitored with a mercury thermometer, and placing the measuring junction in water gradually cooling, for example, from 50 °C to room temperature, also monitored with a mercury thermometer. The water must be stirred. Voltmeter deflections are plotted against temperature at about every 5 °C. The calibration is practically linear over 0–50 °C.

Emphasis is laid on the use of thermocouples for temperature measurements during physiological experiments, because only too often environmental temperatures are stated but the temperature of the material which is the more relevant is not mentioned. All illuminated material and much of respiring material will be at substantially different temperatures from the environment. Temperature control during an experiment should really be based on the temperature of the experimental material. Thermocouples or thermistors are ideally suited for this.

The measurement of leaf temperature (Expts 8.1–3), depression of freezing point (Expt 5.5c) and of xylem translocation rates (Expt 9.1d) are examples of the use of thermocouples. For the measurement of atmospheric vapour densities, wet and dry thermocouples (psychrometers) provide a ready technique.

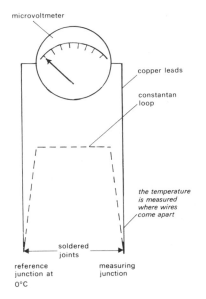

Figure A6 Circuit diagram of copper–constantan thermocouple for measurements of leaf and stem temperatures.

ultramicroscope   see **dark-field microscope**

Experimental material and its preparation

Water

Practically all experiments with plant material involve the use of water and it is necessary therefore to draw attention to the three different kinds of water available in the laboratory. For the preparation of reagent solutions and in all chemical analyses, distilled water must be used. Its pH is usually low and, if this is of importance, it should be noted. However, for many tissues distilled water is deleterious, especially during prolonged treatments, e.g. blank treatments during determinations of tissue water relations or for incubation in Experiments 3.4, 3.5 and 5.4. For such treatments tap water is recommended, unless in a particular area tap water is known to be unsuitable, in which case, if distilled water is used, it should contain 5×10^{-1} mol m^{-3} (0.5 mM) CaCl$_2$.

For keeping plant material collected in the field, tap water should be used; plant material growing in water, e.g. *Elodea*, should be kept in a container filled with that water until used in an experiment.

The use of deionised water is prescribed in some specific procedures, but deionised water is by no means free of solutes and its use can lead to absurd results, especially in amino-acid and allied analyses.

Plant material

COLLECTING AND PREPARING

Cut branches

For physiological experiments plant material must be in sound condition but, if it is cut off, its food and water supply will be disrupted. It is important, therefore, to minimise the damage done. Woody twigs should be cut off in the early morning or late afternoon when water stress can be assumed to be mild. If possible, the cut should be made under water but, since this is often not practicable, the best procedure is to place the cut end of the twig quickly in a bucket or other container of water and to make a second cut about 5 cm from the cut end of the twig under water. If this is not done in the field, it must be done when the material is brought into the laboratory. If water is not available when collecting, it will help to place a plastic bag over the leaves to reduce transpiration and the development of tension in the xylem while the material is transported to the laboratory. These precautions will prevent air locks in the xylem.

When cut twigs are used in potometers or any other assembly, they should first be fitted through the hole in the bung and thereafter the last 3 cm of the bark at the cut end should be removed to prevent phloem exudate gumming up the xylem. On no account should Vaseline be used in the mistaken belief that this will make an airtight fit.

Fresh water and marine material

Chara, Nitella, Chlorella, Scenedesmus, Ulva, Fucus, Porphyra and *Elodea* are used in several experiments (e.g. Expts 3.6, 5.6b & 8.10). The first two are best used fresh after collection, but they can be kept in pond water (see Part C1) for some time. *Elodea*, if convenient, may be collected fresh on the day before use in the laboratory and be kept in its own water. It often requires adhering filamentous algae to be cleaned from it. For *Scenedesmus* and *Chlorella* see below.

Marine algae should be collected and used fresh immediately after collection, but for some experiments (e.g. Expt 8.10) a bulk supply can be kept in a deep freeze. However, *Ulva lactuca* keeps for at least 2 weeks in a generous supply of sea water in proportion to the algal mass and with regular seawater changes. The water should be aerated and kept at 12 °C.

Leaves

Leaves collected for experiments should be excised with a wet razor blade and kept turgid either by placing their petioles in water and their laminae under a plastic hood or by floating their lower surfaces on water until used.

Epidermal peels

It is essential that the plants from which the leaves are taken are grown as described below, so that they will not have been exposed to water stress and will be in the correct phase of their rhythm. The procedure for peeling off epidermal tissue should be standardised. Leaf tissue strips should be devoid of major veins and about 0.5 cm wide. These strips can be placed on a wet microscope slide or over an index finger with the upper epidermis facing up. With a sharp new blade an incision is then made through the upper epidermis and into the underlying mesophyll tissue, leaving the lower mesophyll tissue and the lower epidermis intact. The tissue strip should now be placed on a microscope slide with the upper epidermis facing down. The lower epidermis is now facing the experimenter who can lift a tab of tissue with a good pair of forceps and bend it back sharply as shown in Figure A7. While pulling the epidermis free of the other tissue, the leaf tissue should be held on to the slide with two fingers or with another pair of forceps and the tab pulled vertically upwards. The lower epidermis will come readily off the remainder of the tissue. The angle between the epidermal peel and the remainder of the tissue should be obtuse or a right angle because, if the epidermis is pulled back at an acute angle, damage to many epidermal cells will occur. In some species epidermal cells will be damaged in any case: in *Commelina communis* 30% will be damaged; in *Vicia faba* practically only the guard cells will remain functional.

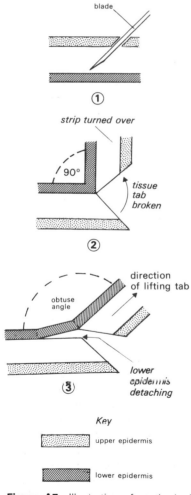

blade

①

strip turned over

90°

tissue tab broken

②

direction of lifting tab

obtuse angle

lower epidermis detaching

③

Key

upper epidermis

lower epidermis

Figure A7 Illustration of method of peeling epidermal tissue from turgid leaf tissue.

If the purpose of peeling the epidermis off the leaf tissue is to observe the degree of stomatal opening, it is best to place the leaf strip in a drop of liquid paraffin (sp. gr. 0.86) while peeling off the epidermis; this will preserve the degree of opening whereas, if peels are taken in water, this tends to cause partial stomatal closure. Peels prepared in liquid paraffin should be mounted in liquid paraffin – it makes microscopic observation very clear. Immersion of the leaf strip in liquid paraffin during peeling has another advantage: it prevents the curling up of the strip that often occurs when peeling is carried out in water. If the latter is necessary, it is good practice to leave a piece of green tissue 0.2–0.3 cm wide at either end of the strip to prevent it curling up.

Many intact leaf tissues mounted in liquid paraffin permit the satisfactory observation and measurement of stomatal pore widths, making the peeling superfluous. However, for Experiments 2.9, 2.10, 3.2, 3.3, 6.4 and 6.5 epidermal peels taken in water are essential. Incubation media for epidermal strips are specified in the particular experiments, but as a general rule it should be routine to add 5×10^{-1} mol m^{-3} (0.5 mM) CaCl$_2$ to the medium as it is generally agreed that this preserves 'membrane integrity'.

Hydathodes

These specialised vein endings are common in the margins of almost all dicotyledonous leaves and the tips of cereal leaves (cf. Expt 5.1). Individual serrations, cut off at their base, should be placed in 10 cm^3 (ml) lactophenol (see Part C1) in a 50 cm^3 (ml) beaker. If left at room temperature, the tissue will 'clear' in 1–2 days, but repeated gentle heating over a Bunsen flame is recommended until the liquid begins to vaporise, followed by cooling, swirling and reheating a few times. The tissue will clear speedily. It is best to do this on the day before the practical class and leave the cleared tissue standing overnight in lactophenol or mounted in lactophenol on microscope slides with coverslips. Tissue should be handled with camel-hair brushes. This procedure is also recommended for the anatomical study in Experiment 11.13.

POLLEN GRAINS

The growth of pollen tubes is a sight every plant physiologist should observe. With many species the process is fast enough to be quantitatively studied in a 3 h practical class. Such measurements provide a useful opportunity to practise microscopic measurements. Pollen grains must be fresh from the flower, but it is impossible to be certain when the pollen is in the 'best' state for germination. Only trials with many flowers from the same and different species will identify the 'best' pollen grains. Microscope slides with a drop of medium (see below) should be prepared and the anthers either dipped into the medium or held above the microscope slide and 'dusted' off with a camel-hair brush. It is best to use several anthers to obtain a mixture. The dusting method is better, as clumps of pollen grains are not desirable; they should be spread out.

The success with different species will depend on season and locality; however, glasshouse-grown *Vicia faba* can be recommended. A winter-flowering glasshouse plant, *Bilbergia* (cultivated spp.) produces pollen grains over 3 weeks that germinate readily. Most spring flowers and fruit trees provide good material. Fresh pollen grains of *Crocus* spp. can be placed on stigmatic surfaces of excised styles which have been split longitudinally to allow the microscopic observation of the pollen tubes growing down the styles. The preparation must be kept in a moist atmosphere.

For studies of the exine sculpture, pollen grains collected from *Helianthus annuus*, *Zea mays*, *Lilium* spp. and, indeed, most other flowers can be kept dry in stoppered sample tubes for longer than 2 years.

Medium

A very reliable medium for pollen-grain germination can be made from 3 g Oxoid (see Part C2) Czapek Dox agar plus 5 g sucrose in 100 cm³ water.

GROWING PLANTS IN SOIL

The plants listed below are a small selection. In different localities and climatic regions other species may well be used, but they should be grown from seed or cuttings in uniform environments and prevented from developing water stress, preferably by a form of automatic watering. The simplest method is to place all pots in a gravel bed through which there is a continuous flow of water.

For some experiments, e.g. Experiments 6.1–6.5, diurnal rhythms are important and these may not fit in with timetabled practical classes. In this case 10 days before a class meeting that is to start for instance at 14.00 h, plants should be transferred to a suitable light–dark regime, e.g. illumination beginning at 12.00 h (noon) so that plants will be in their morning phase in the afternoon.

PLANT SPECIES FOR PHYSIOLOGICAL WORK MENTIONED IN THE TEXT

Species mentioned here are those which can be grown in botanical gardens and glasshouses. Species that must be collected from their respective habitats are not included and those which must be cultured are dealt with below.

Grown from seed

Amaranthus caudatus
Avena sativa (oat)
Beta vulgaris (beetroot)
Brassica alba (white mustard)
Brassica oleracea (cabbage)
Commelina communis
Cucumis sativus (squash)
Cucurbita pepo (cucumber)
Cyclamen persicum
Helianthus annuus (sunflower)
Hordeum vulgare (barley)
Lactuca sativa (lettuce)
Lepidium sativum (cress)
Lolium perenne (rye grass)
Lupinus alba (lupin)

Lycopersicon esculentum Mill. (tomato)
Nicotiana tabacum (tobacco)
Oryza sativa (rice)
Phacelia tanacetifolia
Phaseolus aureus (mung bean)
Phaseolus vulgaris (dwarf or runner bean)
Ricinus communis (castor bean)
Spinacia oleracea (spinach)
Triticum vulgare (wheat)
Xanthium strumarium (cockle burr)
Zea mays (maize)

Grown from stem cuttings

Coleus blumei   *Pelargonium zonale*
Hedera helix (ivy)   *Rhoeo discolor* (boat plant)
Impatiens sultani (busy Lizzy)   *Tradescantia virginiana*

Grown from bulbs, stolons or root cuttings

Bilbergia spp. (cultivated)   *Lilium* spp.
Crocus spp. (cultivated)   *Saxifraga tomentosa*

GROWING IN LIQUID MEDIA

The only species recommended for growing in water is *Elodea canaden-sis*. After collecting the plant in a bucket filled with the water in which it grows, it should be cleaned by removing adhering filamentous algae and then transferred with its water to vats sunk into the ground so that their rims are 3–5 cm above the soil. The volume of water can be made up with tap water and in most localities rain will replenish the water lost by evaporation. The plants usually thrive; ice formation on the surface of the vats in winter does not harm the plants.

Anabaena, *Nostoc*, *Chlorella*, *Scenedesmus* and *Saccharomyces*, are cultured in special solutions and under laboratory conditions.

BLUE-GREEN ALGAE: *ANABAENA AND NOSTOC*

(Based on information given by Professsor W. D. P. Steward and Miss G. Alexander, University of Dundee.)

Suitable strains are available from the Culture Centre of Algae and Protozoa (see Part C2): *Anabaena cylindrica*, cat. no. CCAP 1403/2a; *Nostoc ellipsosporum*, cat. no. CCAP 1453/2. These cultures are not axenic; axenic cultures may be obtained from the American Type Culture Collection (see Part C2). These cultures are much more expensive.

The following composition of the culture solution (BG11) is taken from Stanier *et al.* (1971).

$NaNO_3$	1.5 g	EDTA (disodium magnesium)	0.001 g
K_2HPO_4	0.04 g	Na_2CO_3	0.02 g
$MgSO_4 \cdot 7H_2O$	0.075 g	trace metal mixture	1.0 cm^3
$CaCl_2 \cdot 2H_2O$	0.036 g	agar (if needed)	10.0 g
citric acid	0.006 g	distilled H_2O	1000 cm^3
Ferric ammonium citrate	0.006 g		

Trace metal mixture:

H_3BO_3	2.86 g	$CuSO_4 \cdot 5H_2O$	0.079 g
$MnCl_2 \cdot 4H_2O$	1.81 g	$Co(NO_3)_2 \cdot 6H_2O$	0.049 g
$ZnSO_4 \cdot 7H_2O$	0.222 g	distilled H_2O	1000 cm^3
$NA_2MoO_4 \cdot 2H_2O$	0.39 g		

After sterilisation, the pH of the medium should be 7.1; it should not be allowed to become acid; if necessary adjustment with KOH is recommended.

For nitrogen-fixing algae $NaNO_3$ can be omitted and the cobalt nitrate replaced with another cobalt salt.

The temperature of the culture should be between 20 and 25 °C, illuminated at about 80 μmol m^{-2} s^{-1} (80 μE m^{-2} s^{-1}). Cultures up to 200 cm^3 can be shaken to provide aeration: larger volumes should be aerated with sterile air.

Cultures should be inoculated with sufficient cells to make them appear green to the eye (about 0.3 μg chlorophyll a cm^{-3}); if the culture is too thin, 'bleaching' may occur with subsequent failure to grow. A log phase culture will have 2–3 μg chlorophyll a cm^{-3} and would have a C$_2$H$_2$ reduction rate (see Expt 2.8) of about 5–20 nmol (10^{-9} M) C$_2$H$_2$ reduced per μg chlorophyll a h^{-1}.

CHLORELLA

(Based on information given by Dr J. Hannay, Imperial College of Science and Technology.)

Two stock solutions are prepared and finally mixed in the proportion 2000 cm^3 of solution A and 4 cm^3 of solution B.

Solution A
10 cm^3 10^3 mol m^{-3} (1.0 M) KNO$_3$
 1 cm^3 10^3 mol m^{-3} (1.0 M) K$_2$HPO$_4$
 4 cm^3 10^3 mol m^{-3} (1.0 M) MgSO$_4$
 1 cm^3 10^3 mol m^{-3} (1.0 M) KH$_2$PO$_4$
 0.5 cm^3 10^3 mol m^{-3} (1.0 M)
 Ca(NO$_3$)$_2$·4H$_2$O
 2 cm^3 10^3 mol m^{-3} (1.0 M) KCl

made up to 2000 cm^3

Solution B
11.0 g ZnSO$_4$·7H$_2$O
5.7 g H$_3$BO$_3$
2.2 g MnSO$_4$·H$_2$O
2.5 g FeSO$_4$·7H$_2$O
0.84 g Co(CH$_3$COO)$_2$·4H$_2$O
0.78 g CuSO$_4$·5H$_2$O
0.55 g (NH$_4$)$_6$Mo$_7$O$_{24}$
25.0 g EDTA disodium
made up to 500 cm^3

It is recommended to grow two cultures in 500-cm^3 Dreschel bottles arranged as shown in Figure A8. Aeration must be vigorous to prevent

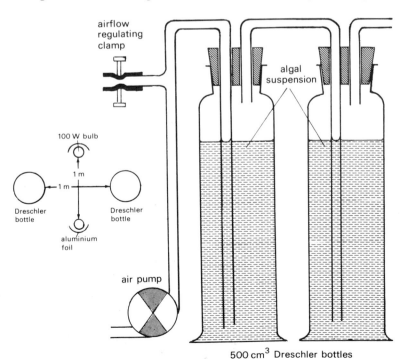

Figure A8 Arrangement of apparatus seen in longitudinal section for the preparation of *Chlorella* cultures. The position of the culture vessels relative to the two electric lamps is indicated in plan.

airflow regulating clamp

algal suspension

100 W bulb

1 m

1 m

Dreschler bottle

Dreschler bottle

aluminium foil

air pump

500 cm^3 Dreschler bottles

Chlorella cells from settling out. Continuous illumination can be provided by two 100 W lamps 50 cm distant from the bottles. The pH of the culture should be kept between 6.5 and 7.0, and the temperature at 25 °C; if necessary, the lamps may be cooled by a fan.

The recommended system of two culture bottles is operated by inoculating one bottle with about 10% of the 'old' culture into fresh medium. The 'old' culture should be ready for harvesting in between 5 and 7 days and should appear deep green. If necessary, the culture could be centrifuged and resuspended in fresh medium to concentrate it. There should be about 10 mm^3 cells cm^{-3}, i.e. about 10^6 cells cm^{-3}, each cell with a diameter of about 10 μm.

Scenedesmus obliquus

(Based on information given by Dr K. Hardwick, University of Liverpool.)

Cultures for inoculation of the medium can be obtained from the addresses cited for blue-green algae earlier in this section.

Three basic stock solutions are required as shown below:

> solution A   202.0 g KNO$_3$
> 117.0 g MgSO$_4$
> 62.0 g NaCl
> 0.1 g ZnSO$_4$
>
> *made up to* 1000 cm^3;

> solution B   75.0 g CaCl$_2$ in 1000 cm^3 water

> solution C   151.0 g Na$_2$HPO$_4$ dissolved in 500 cm^3 water and the pH adjusted to 6.5 with HCl before making up to 1000 cm^3

Solutions A, B and C are used in the preparation of 5000 cm^3 of the final culture solution as follows:

> 20.0 cm^3 solution A
> 1.0 cm^3 solution B
> 20.0 cm^3 solution C
> 0.05 g FeSO$_4$
> 0.001 g MnSO$_4$
> 25.0 g glucose
> 12.5 g Difco (see Part C2) yeast extract
>
> *made up to* 5000 cm^3

The temperature of the culture solution should be kept between 25 and 30 °C and its pH at 6.5. Illumination is continuous at between 20 and 100 μmol m^{-2} s^{-1} (100 μE m^{-2} s^{-1}) photosynthetic active radiation, but does not seem to be critical. The culture should not be aerated but continuously shaken, preferably on a rotary shaker. The inoculum should be distinctly green before incubation in order to produce large volumes of suspension. About 80 cm^3 culture are grown in 250-cm^3 conical flasks which should be plugged with cotton wool. Sterile procedures should be followed for inoculation. The cells seem to be photosynthetically most active 2 days after inoculation but remain active for up to 14 days.

SACCHAROMYCES SPP

The species recommended for Experiment 1.8 is commercially available baker's yeast as supplied by health food shops and bakeries. To prepare this yeast for use, 5 g are suspended in 100 cm³ water and aerated at room temperature for 24 h. If other yeast species are used, these must be grown as cultures to yield an adequate weight after centrifugation.

SACCHAROMYCES CERVISIAE

This species is grown in 0.2 g glucose dissolved in 100 cm³ water. The culture may be shaken or aerated at 20 °C. Experimental suspensions are prepared by using 5 g of centrifuged yeast resuspended in 100 cm³ water kept aerated at room temperature for 24 h.

REFERENCES

Fogg, G. E., W. D. P. Stewart, P. Fay and A. E. Walsby 1973. *The blue-green algae.* London: Academic Press.

Stanier, R. Y., R. Kisana, M. Mandel and G. Cohen-Bazire 1971. *Bacteriol. Rev.* **35**, 171–205.

ISOLATING ORGANELLES

CHLOROPLASTS

(Based on information given by Dr R. Phillips, University of Stirling.)

Most commonly used leaves are lettuce (*Lactuca sativa*) and spinach (*Spinacia oleracea*). They are kept in a tray filled with tap water before use to keep them turgid. About 6 g fresh wt of lamina tissue without major veins is cut into pieces of about 10 cm² and macerated in either a mortar or a glass blender together with an initial 15 cm³ suspension medium (see below), gradually increasing the volume by adding more of the medium until 30 cm³ have been used.

The ground-up tissue *brei* is filtered and squeezed through four layers of muslin into an ice-cold beaker. After centrifuging the filtrate at 3000 g for 5 min, the supernatant is decanted and the pellet resuspended in 3.0 cm³ suspension medium using either a Whirlimixer from Fisons (see Part C2) or a camel-hair brush. This chloroplast suspension must be kept ice-cold.

Suspension medium

The recommended suspension medium is made up as follows:

$$10 \text{ cm}^3 \ 10^3 \text{ mol m}^{-3} \ (1.0 \text{ M}) \text{ sorbitol}$$
$$7.5 \text{ cm}^3 \ 10^2 \text{ mol m}^{-3} \ (0.1 \text{ M}) \text{ K}_2\text{HPO}_4$$
$$6.0 \text{ cm}^3 \ 25 \text{ mol m}^{-3} \ (25 \text{ mM}) \text{ NaEDTA}$$
$$6.0 \text{ cm}^3 \ 10 \text{ mol m}^{-3} \ (10 \text{ mM}) \text{ NaHCO}_3$$
$$0.5 \text{ cm}^3 \ \text{H}_2\text{O}$$

The final volume is 30 cm³ (ml). The medium is adjusted to pH 7.6 with KOH or HCl and chilled on ice to near 0 °C.

MITOCHONDRIA

(Based on information given by Dr J. M. Palmer, Imperial College of Science and Technology.)

The materials recommended for the isolation of mitochondria are the tubers of potato (*Solanum tuberosum*) and Jerusalem artichoke (*Helianthus tuberosum*). The latter has the advantage that it is not rich in starch grains (see below).

The tubers should be washed in tap water and peeled, then rinsed in distilled water and packed in ice to cool down to 4 °C. Roughly 0.1 mg mitochondrial protein can be obtained from 1.0 g tuber tissue. All isolation procedures must be carried out close to 4 °C.

In order to grind the tissue, it is best to cut it first into 1-cm³ blocks. The proportion of tissue to extraction medium (see below) should be 2 : 3, i.e. 200 g tissue will require 300 cm³ medium. The tissue should be ground in a top-driven grinder until it appears like porridge. It is then strained through three layers of muslin. The liquid coming through the cloth is next centrifuged at 1000 g for 5 min to remove starch grains – this is especially necessary when potato tissue is used.

The supernatant must now be centrifuged for approximately 120 000 g min, i.e. at 40 000 g for 3 min or at 12 000 g for 10 min.

The pellet is resuspended in 5 cm³ of washing medium (see below) using a glass rod, transferred to a glass Teflon homogeniser and gently homogenised – however, some material will not break up and it is better to under-homogenise at this step than to overdo it. The homogenised suspension in the 5 cm³ washing medium is next diluted with washing medium to 50 cm³ and centrifuged at 1000 g for 5 min to remove any remaining starch grains and cell debris. The supernatant is finally centrifuged at 100 000 g min.

The final pellet is resuspended in a 0.5 cm³ washing medium by means of a Pasteur pipette, using a jet of medium to wash the pellet off the walls of the tube. Any dark coloured material adhering to the tube should be left, as it probably contains peroxisomes. Good mitochondrial material breaks up readily in the Pasteur pipette; if it appears very yellow, it may contain amyloplast membranes from starch grains.

The finished mitochondrial preparation can be kept in a narrow, stoppered tube on ice until required; it will remain active for at least 6 h. A yield of 1 cm³ suspension containing between 15 and 20 mg protein should be obtained from 200 g tissue.

A respiratory assay in an oxygen electrode will require 0.5–1.0 mg protein cm⁻³ suspension.

Extraction medium

The extraction medium is composed of:

5×10^2 mol m⁻³ (0.5 M) sucrose
10 mol m⁻³ (10 mM) MOPS (see Buffers, 'Good', Part C1)
5 mol m⁻³ (5 mM) EDTA

This medium is at a pH of 7.8; it can be stored at 4 °C for several days. Immediately before use, an anti-oxidant is added: 2 mol m⁻³ (2 mM) $Na_2S_2O_5$ together with 0.1 g bovine serum albumin (BSA) in 100 cm³ water.

It will be necessary to readjust the pH to 7.8 *at* 4 °C.

If a bottom-driven grinder is used (e.g. a Waring blender), the aim should be to disrupt only about three-quarters of the tissue cubes.

It is strongly recommended to prepare at least 300 cm³ each of extraction and washing medium.

Washing medium
The washing medium consists of:

4×10^2 mol m^{-3} (0.4 M) sucrose
5 mol m^{-3} (5 mM) TES at pH 7.2 (see Buffers, 'Good', Part C1)

Immediately before use, 0.1 g BSA in 100 cm^3 water is added.

PROTOPLASTS
(Based on information given by Dr J. D. B. Weyers and Mr P. J. Fitzsimons, University of Dundee.)

Purification of cell wall degrading enzyme
Cellulysin 0.1 g; from Calbiochem (see Part C2) dissolved in 5 cm^3 distilled water, is placed in a 1-cm-diameter visking tube from Fisons (see Part C2) for 16 h at 4 °C to allow for dialysis into 1000 cm^3 (1.0 litre) distilled water at pH 8.0, adjusted with KOH. Enzyme batches vary in their activity and it may be necessary to use up to twice the recommended strength. The dialysing purification is not essential unless non-viable and unspherical protoplasts are obtained.

> pH control of distilled water during dialysis is essential as otherwise the visking tube will be degraded.

Preparation of epidermal tissue
Epidermal peels of *Commelina communis* are prepared as outlined in Part A and floated with the cuticle facing up on 5×10^{-1} mol m^{-3} (0.5 mM) CaCl$_2$ in a 15-cm-diameter Petri dish. When about 75 cm^2 epidermal tissue have been collected, the material is transferred to two 9-cm-diameter Petri dishes containing 3×10^2 mol m^{-3} (0.3 M) mannitol and kept there for 30 min in order to allow the epidermal cells to plasmolyse. Collecting the epidermal tissue first in CaCl$_2$ solution is optional; placing the peels directly into the osmoticum may give equally good results.

> It is recommended to prepare 75 cm^2 epidermal tissue to begin with as some will be lost during handling.

Preparation of enzyme digestion medium (EDM)
The 5 cm^3 purified cellulysin is diluted with 5 cm^3 of 6×10^2 mol m^{-3} (0.6 M) mannitol containing 20 mol m^{-3} (20 mM) MES buffer (see Buffers, 'Good', Part C1) at pH 5.0 and 0.05 g bovine serum albumin (BSA) fraction IV from Sigma (see Part C2) so that the resulting 10 cm^3 EDM contains 0.1 g purified cellulysin, 3×10^2 mol m^{-3} (0.3 M) mannitol, 10 mol m^{-3} (10 mM) MES and 0.05 g BSA. If necessary, the pH should be adjusted to 5.0 with KOH.

Digestion of cell walls
About 50 cm^2 of the partially plasmolysed epidermal tissue is floated on 10 cm^3 EDM in a 9-cm-diameter Petri dish so that the medium is covered with tissue. The dish is next sealed with tape and floated on a water-bath at 30 °C for 4–6 h, being swirled around by the water-bath agitator.

Protoplast harvest
After the 4 h incubation the Petri dish is removed from the bath and swirled gently to dislodge naked guard-cell protoplasts from the rest of the tissue. After opening the Petri dish, the EDM, which also contains epidermal-cell and subsidiary-cell protoplasts, is transferred via a

wide-mouth Pasteur pipette into four 15 cm³ polypropylene centrifuge tubes. The tissue remaining in the Petri-dish should be washed with a further 15 cm³ (ml) of 3×10^2 mol m⁻³ (0.3 M) mannitol and the washings added to the tubes. Centrifugation follows at 100 **g** for 5 min. Cell debris remains on the surface of the supernatant and can be discarded together with the supernatant. The pellets from the four tubes are next combined for washing and centrifuging 3 times with 3×10^2 mol m⁻³ (0.3 M) mannitol. The final pellet is resuspended in 1 cm³ (ml) of 3×10^{-2} mol m⁻³ (0.3 M) mannitol and allowed to equilibrate in a covered test-tube kept at 30 °C for 15 min.

Examination of the protoplasts

Guard-cell, subsidiary-cell and epidermal-cell protoplasts will differ in their size and cell content; also in their response to incubation in solutions of different molarities, in which they will swell or shrink. Observation should be carried out on haemocytometer slides as coverslips will otherwise tend to squash the delicate protoplasts. Measurements of protoplast diameters in different media can be carried out.

REFERENCE

Fitzsimons, P. J. and J. D. B. Weyers 1983. *J. Exp. Bot.* **34**, 49–60.

DICHROIC STAINING

(Based on information given by Mr P. St J. Edwards, University of Salford.)

An appreciation of the major direction and orientation of cellulose microfibrils in plant cell walls contributes to an understanding of several cell functions such as water storage, tissue pressure, guard-cell deformations and fruit dehiscence. Fluctuations in the depth of colour seen in cell walls treated with a dichroic stain and viewed in polarised light enable the observer to deduce the substructure of the wall. A convenient stain for this purpose is aqueous Congo red at a concentration of 0.5–1.0 g in 100 cm³ water.

Microscopes with built-in illumination can be fitted with a single polarising filter between the lamp and the condenser. A commercial polaroid filter can be used or sheets of polaroid plastic filters obtainable from Griffin & George Ltd (see Part C2). When this substage filter is rotated, the depth of red stain in certain cell walls is seen to fluctuate more or less markedly during each 180° of rotation:

(a)   The amplitude of this fluctuation indicates the orderliness of the texture of the wall. When cellulose microfibrils are virtually parallel, the colour will seem to disappear twice during each complete revolution of the polariser.

(b)   The plane of polarisation for maximum depth of red colour indicates the predominant direction of microfibrils in a single cell wall or the resultant direction when two cell walls are superimposed.

These two points can be visualised in relation to *delignified* xylem. Figure A9 shows an extended proto xylem helix. Portions of the helix that

Figure A9 Diagram of two extended protoxylem helices as they appear after staining with Congo red in polarised light. Portions of the helices which lie perpendicular to the plane of polarisation appear almost colourless; those which lie in the plane of polarisation appear deep red. In macerated metaxylem vessels the contrasts are less strong as the dichroism of the farther wall interferes with that of the nearer wall. Macerated poplar leaf skeleton is suitable for this exercise.

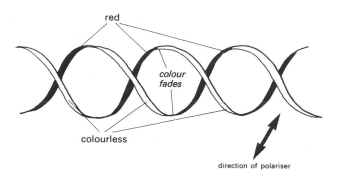

lie perpendicularly to the plane of polarisation seem colourless, whereas aligned portions appear deep red. This indicates that cellulose microfibrils within a helix lie truly parallel to each other. By contrast, macerated *meta*xylem vessels display a less extreme change in depth of stain when the polariser is rotated through 90° because the oblique dichroism of stain in the nearer wall detracts from that of the farther wall. Maximal depth of colour is seen with the polariser plane perpendicular to the long axis of the cell. This is the resultant of two dichroisms superimposed in a helix of shallow pitch.

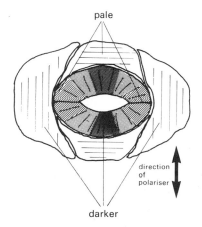

READILY STAINED MATERIAL

Collenchyma found in longitudinal sections of aerial roots of *Monstera delicosa*, petioles of celery (*Apium graviolens*) and of other ribbed stems or petioles can be stained with Congo red in 1 min. Material must be rinsed after staining and mounted in water; if the sections are to be kept for more than 10 min, the coverslip should be sealed with rubber solution or nail varnish. Cell-wall thickening will appear dark red when parallel to the plane of polarisation and colourless when at 90° to it. This indicates that the cellulose microfibrils are orientated longitudinally to the axis of the cell so that the cell can become narrower, shortening when the tissue loses water.

   This interpretation, as all others following, is based on the view that *cell wall extensibility is greatest at right angles to the major orientation of the microfibrils* (Ziegenspeck 1938).

Figure A10 Ellipse-shaped and graminaceous guard cells with radiating and parallel microfibrils and their surrounding subsidiary cells showing contrasting depth of colour in polarised light after staining with Congo red.

SIMMER STAINING

(a)   Epidermal strips (see Part A) of *Tradescantia virginiana*, *Zebrina pendula* and *Pelargonium zonale* with radial microfibrils in guard-cell walls, and *Zea mays* with parallel microfibrils (see Fig. A10) are suitable material. However, observation of stained material is not as clear as the microscopic picture of unstained material in a polarising microscope with analyser – an observation that is strongly recommended. For staining, pieces of epidermis are submerged in a shallow volume of stain in a suitable container so that it can be brought to the boil and simmered for not less than 2 min. The addition of a trace of a neutral wetting agent will help to submerge the tissue. After thorough rinsing with water, the

The direction of polarisation should be marked on the filter mounting. In order to do this one must look through the filter against the light at a horizontal shiny surface. The position of the filter, after rotating it until all glare has disappeared, will allow only vertically polarised light to pass through.

The single polarising filter should be mounted in a rotatable ring held in a retort-stand clamp so that the filter can be swung in between light source and condenser.

Since reflection from a mirror at 45° causes polarisation, a light beam from a separate microscope lamp reflected into the condenser by a mirror will produce a flicker when the polarising filter (which now becomes a second polariser) is rotated. Hence, microscopes with built-in illumination without a mirror are preferable for observing material treated with a dichroic stain.

Geological polarising microscopes can be used with the analyser withdrawn from the light path in order to observe the effect of dichroic staining. But, if such a microscope is available, the unstained material observed with both polariser and analyser will show the microfibrillar patterns very well.

If, after the addition of NaOH and standing overnight, the macerate appears to be acidic again, some more NaOH must be added.

Water is specified as the mountant because glycerol diminishes the dichroism of Congo red staining. Glycerol is therefore unsuitable for the preservation of macerates, and water-rinsed macerates should be dried on a paper towel, packeted and stored in the dark until needed. To reconstitute dried material, it is soaked in water in a test-tube and brought to the boil to dispel air. After cooling, the suspension should be shaken to disperse the macerate which, with the addition of a little more stain if necessary, can be mounted and examined as described above.

epidermal tissue is mounted in water under a coverslip. When the polariser is rotated continuously, the staining pattern will appear to rotate also, indicating a radial arrangement of the microfibrils in the guard-cell walls; however, in those of *Z. mays* the microfibrils are parallel to the length of the cell, except in the tiny portions at either end.

(b) Pieces of dandelion peduncle (*Taraxacum officinale*), 1.0 cm long, fresh or preserved, and preferably at different stages of growth are treated and stained as in (a) and then slit open. The inner tissue is scraped and teased out in water on a microscope slide, and is then spread out under a coverslip for observation. Some epidermal tissue of the peduncle can be mounted separately. The parenchyma cell walls will be seen to have transverse dichroism, whereas the epidermal cells and collenchymatous cells have longitudinal dichroism.

MACERATED TISSUE

The valves of mature, fresh or preserved (in 70% ethanol) legume pods are good material to work with. *Lathyrus latifolius* is best, but garden pea is suitable also. The outer, green, soft mesocarp tissue should be scraped off first, leaving the fibrous endocarp, which should be shredded and then macerated by refluxing overnight in equal parts of glacial acetic acid and 100 vol. hydrogen peroxide (see Part C1). After cooling, the bleached macerate should be diluted with water and strained through nylon cloth (coffee strainers have been found effective) and finally rinsed thoroughly with water. Staining with Congo red is carried out in a small beaker. The stain will at first blacken, but by gradually adding 10^3 mol m^{-3} (1.0 M) NaOH and stirring until all acidity has disappeared, the stain will become red. Time should be allowed for acidity to reappear as lumps of tissue separate and then more NaOH should be added. The preparation must be strained once more and after rinsing is suspended in water.

The two kinds of fibres that will be seen are 'scalloped' and 'plain'. Both kinds take up stain, but transverse dichroism will be seen only in the scalloped fibres. As a result of the different orientation of the microfibrils, the plain fibres shrink laterally, but this shrinkage is opposed by the adjacent layer of scalloped fibre and crystal-containing sclereids. This causes the fibrous endocarp layer to curl and the pod to dehisce.

This study can be complemented by a demonstration of dehiscence using ripe, but unopened, pods of *Lathyrus latifolius* or *Lupin* spp., which should be stored in the dark above a saturated sodium nitrite paste in sealed Kilner jars. Unripe fruit should be avoided as they may become mouldy. While drying the fruit in front of an electric bar heater, dehiscence will occur, scattering the seed over a considerable distance.

REFERENCE

Ziegenspeck, H. 1938. *Bot. Arch.* **39**, 268–309; 332–72.

Experimental reports

The objective of experiments is to test hypotheses; for this purpose, measurements are made under controlled conditions and the results are recorded in tables or graphically presented. On the basis of such records, conclusions can be reached as to the validity of a specific hypothesis. Quantitative data often need to be statistically analysed. It is therefore necessary that experiments are designed with the analysis in mind. With variable plant material, the need for this can be demonstrated in class experiments when results from several groups are pooled in order to arrive at valid conclusions. Statistical design and analysis of results are not arithmetical exercises for their own sake but in many cases the only means of testing the validity of hypotheses.

Results must also be presented in the most cogent form and it is in this context that headings of reports and captions for tables and figures, and the specific form of the latter, are important. They must reflect a correct attitude to experimentation in general. For instance, captions beginning 'Experiment to prove . . .' or 'Graph to show . . .' and plotted points connected by freehand curves are indicative of preformed views of what the results ought to be. This view of experiments, as confirmation of results predicted by authority, may have its place at some stage in the learning process but, if the educational aim is to stimulate enterprise, the approach should be open minded, somewhat sceptical and critical, but above all on the look-out for the unexpected, for variations and new insights requiring further experiments.

Tables

During and after an experiment measurements must be recorded. This is best done in tabular form and applies to raw data as well as to all kinds of transformed data. The layout of a table should be decided before the experimental measurements begin so that data can be recorded under their respective rubrics. Such an initial layout may prove unsatisfactory when data accumulate and transformations are calculated or the need for additional information becomes apparent. Protocol tables contain the essential information, but for a final report another layout may be preferable.

It is a matter of taste and available space on the page as to which entries are placed across the table and which vertically, but one arrangement can be better than the other for ease of following comparisons between results of different treatments or of different species. Where subdivided rubrics are needed, these are best at the head of the vertical columns.

Once the data are finally tabulated, it is essential that the table is adequately captioned, irrespective of whether it is for class use or for

Table x Changes in temperature of dry starch on adsorption of water. (Initial temperature of dry starch: 18 °C; initial temperature of water: 16 °C.)

Number of drops of water added	Change in temperature of mixture (°C)
3	+ 4.0
6	+ 3.0
9	+ 1.5
12	+ 0.5
15	− 0.7

Because both the addition of water and the stirring must be carried out at reasonable speed, it is not practicable to measure the quantity of water added on each occasion accurately by volume — hence the use of 'drops' as a measure.

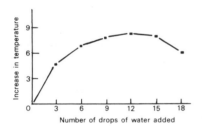

Figure x Change in temperature of dry starch on progressive adsorption of water. (Initial temperature of starch: 18 °C; initial temperature of water: 16 °C.)

publication. Without an adequate caption, a table of results can be completely meaningless to a reader. Concise and lucid captions are not easy to formulate and few people are able to do so at a first attempt. Although brevity is of great value, a good caption must give enough information for an interested person to gain from the table as a whole the essence of the results of the experiment without necessarily reading the whole report. Time and thought are therefore needed to formulate a caption that fits this requirement, by going over a first draft several times and improving it by additional words or reformulating the wording to make it more concise or more explanatory.

The rubrics of the table also need some thought but are usually more readily formulated. There is almost always one basic requirement: the unit of measurement should be stated in the rubric of each column.

The following example illustrates points to be considered when composing a caption for Table x (or Fig. x).

Table/Figure x   Table/Curve to show increases in temperature

superfluous   *incorrect*

of starch when water is added.

incomplete   *ambiguous*

The next version is more explanatory but long winded:

Table/Figure x   Changing rates of release of kinetic energy in the form of heat during the progressive adsorption of water by dry starch. (Initial temperature of dry starch: 18 °C; initial temperature of water: 16 °C.)

A simpler, but adequate version could be:

Table/Figure x   *Changes* in temperature of *dry* starch on *adsorption* of water. (Initial temperature of dry starch: 18 °C; initial temperature of water: 16 °C.)

By convention, captions for tables are placed above the tables and those for figures always underneath them; where this is not the case in this book, it is for reasons of graphic design.

Graphs

As with tables a concise and lucid caption is essential for the graphical presentation of selected data.

In reports of class experiments both tables and graphs should be prepared. For publication usually either the one or the other will be chosen, but as a training and educational practice in the laboratory both are needed. Thus, a further task of an experimenter reporting on his results is the preparation of the most telling form of graphical presentation. To select the data that are to be presented in graphical form requires an understanding of the physiological significance of the experiment. A decision has to be made as to whether a histogram, a two- or three-dimensional graph or a graph on log paper is required. Frequencies, discrete and discontinuous measurements are suitable for histograms; continuous processes should be presented by curves and, if there are several interacting factors, three-dimensional graphs. Besides theoretical understanding, the preparation of a graph requires technical skills that are, however, accessible to everybody:

(a) The two ruled axes of the graph should be at least 3 or 4 cm in from the graph-paper margin, so that there is space for labels and units for the axes and for a caption underneath the graph. Where graphs represent positive and negative values, the latter must be entered to the left or below a suitably placed zero point as in Figure 5.8c and in Experiment 5.4b.

(b) An appropriate scale to emphasise and make vivid the measured changes must be decided upon for the dependent variable entered on the y-axis and likewise a suitable scale must be chosen for the independent variable on the x-axis (e.g. time, molar concentration, temperature).

(c) Plotted points are best connected with straight ruled lines, leaving a small gap either side of each point, thus: ——·——. The drawing of freehand curves roughly connecting the points can be quite misleading and unrepresentative of the true course of a process. In some cases it is good practice to add one or two labels to individual curves in a single graph: for instance, the names of different species used in an experiment, or the molar concentrations of solutions in which tissue has been immersed while changes in weight are plotted against time (Expt 5.4).

(d) The temptation to extrapolate or to continue a curve to the origin should be resisted unless the necessary assumptions are clearly stated. A frequently occurring example of the folly of extrapolation to the origin occurs when rates of a process against temperature are used for the estimation of temperature coefficients (Q_{10} values) for specific temperature intervals; by implication the rate of the process at 0 °C is assumed to be zero, which is mostly incorrect. If there is scatter of plotted points, these are best left unconnected, unless the experimental procedure allows for a regression analysis resulting in a computed regression line.

Text

The writing of the experimental report follows the tabulating and plotting of results. It must supply sufficient information for someone else to be able to repeat the experiment. Nevertheless, the report is not meant to be an instruction sheet but an account of what has been done. It should therefore be written in the past tense and as much importance should be attached to the precise use of language as to precision in measurement and in the use of numbers.

For example, a report might state: 'The osmotic potential of epidermal cells was measured by plasmolysis.' Such a statement is imprecise because cells do not have an osmotic potential. The relevant term for cells is 'water potential' which is the sum of osmotic and turgor pressure potentials; plasmolysis should be specified. The amended sentence might read: 'The osmotic potential of the vacuolar sap of epidermal cells was estimated by the 50% plasmolysis method.'

Taking another example: 'The compensation point of maize was measured in an infra-red gas analyser' is a faulty statement. It might instead read: 'The *carbon dioxide* compensation point of a maize *leaf* was measured in a *closed circuit* infra-red gas analyser.'

Reports are most readily organised under subheadings without being stereotyped and rigid about them. An *introduction* affords the opportunity for the student to show his understanding of the physiological significance of the experiment and to state what the aim of the experiment is, i.e. what hypothesis is being tested.

The introduction will be followed by a report on the technicalities of the experiment, *the material, apparatus and methods* used. Significant features of the apparatus and the materials should be commented on and, if a complex assembly was used, reference to a diagram is the most informative method of dealing with it (e.g. Expt 7.3). Likewise, special materials, strengths of solutions, energy sources and constant environmental factors should be stated and the selection, preparation and condition of the plant material must be documented. The experimental design, where appropriate, must be defined.

The content of tables and figures should not be repeated in the text but only elaborated in order to highlight salient points. This will lead to a *discussion* of the results. For instance, some feature of the tabulated results or of a graph may well be due to a rhythm in the experimental material, or its age, or a weakness in the experimental procedure only afterwards recognised – indeed, on account of scrutinising the results. Thus a discussion of the data must begin with a critical appraisal of the results.

Whether *results* are combined with a *discussion*, or both are combined with *conclusions*, will depend on the level of study the student has reached and on the context in which the class experiment has been carried out in relation to the lecture course. If the course is an integrated one with ecological topics, clearly laboratory results obtained with certain species can form the basis for widening the discussion to include ecological matters.

Problem-solving exercises

In order to train students to interpret data it is useful to provide exercises using experimental results selected from the literature. These can be in the form of tables and graphs, together with information about the materials and methods used. Such exercises are akin to problem-solving practised in other disciplines. It will be necessary to interpret the results utilising all the information given in the exercise, including experimental design and procedures. If this ability to interpret information is mastered, it will be of very general application and a great help in the writing of laboratory reports.

Part B   Experiments

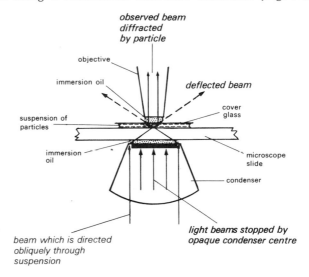

1 Properties of matter

The processes illustrated by the experiments in this section have direct relevance to many experiments in the other sections, especially sections 2, 3, 4 and 5.

KINETIC ENERGY

As all physiological processes derive some of their energy from the kinetic energy of matter, a few experiments dealing with this most basic property of inanimate matter provide a good starting point for practical class exercises in plant physiology. The experiments are not complex and some offer a good introduction to tabulating and presenting graphically the results obtained as well as the opportunity for succinct discussion in experimental reports.

1.1 Brownian movement

To dilute the Indian ink it is best to place a drop of water on a microscope slide and touch it with a rod of about 2 mm diameter that has been dipped into the ink. Once the coverslip has been placed, the appearance should be brown not black.

This basic microscopic observation of one manifestation of the kinetic energy of matter can often be made in cells of plant tissue sections or epidermal peels under 400× magnification, but it is best made using a dark-field or ultramicroscope (see Part A), either with diluted Indian ink or better in a suspension of ground glass (see Part C1).

When using a commercial dark-field condenser (Fig. B1.1), it is

Figure B1.1 Longitudinal section of dark-field condenser with paths of light for the observation of Brownian movement.

essential to centre it accurately and to place glycerol or immersion oil (depending on the condenser) between it and the underside of the microscope slide, as well as between the coverslip and the objective if immersion objectives are used.

If demonstrations are set up, it will be necessary from time to time to add a drop of water at the edge of the coverslip to prevent drying; but observers must not mistake the resulting flow of particles across the field for Brownian movement – they must wait till the flow has ceased.

1.2   Diffusion, gaseous

This consequence of the kinetic energy of matter is most spectacularly demonstrated in the gaseous state by using a porous pot connected to a long U-tube manometer filled with manometer liquid (see Part C1 and Fig. B1.2).

When an inverted beaker filled with hydrogen from a Kipp's apparatus or from a cylinder is placed over the porous pot, the pressure will rise instantaneously at great speed, as the small H_2 molecules diffuse much faster into the pot than the N_2, O_2 and CO_2 molecules can escape from it, thus creating a considerable pressure inside the pot.

When the manometer liquid approaches its limit, the inverted beaker must be removed quickly but kept inverted; the liquid will fall immediately and the inverted beaker should be replaced over the porous pot before the manometer liquid can enter the porous pot.

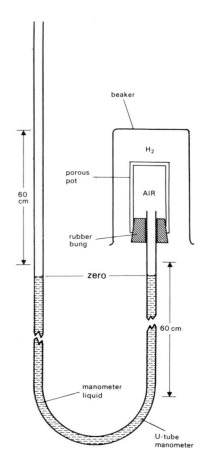

Figure B1.2 Longitudinal section of U-tube manometer with porous pot attached for the demonstration of gas diffusion.

1.3   Diffusion in liquid and colloidal states

During diffusion in a liquid medium, thermal convection currents occur which are difficult to exclude in class experiments. For this reason diffusion in colloidal gels is preferable; the rate of diffusion is hardly less rapid in a gel than in a liquid.

A colloidal gel is made from 3 g gelatine in 100 cm³ water which is poured into a set of test-tubes up to their rims and then allowed to set at room temperature. After the gel has formed, each tube must be made up to a convex meniscus with a few drops of the sol because the volume will have shrunk a little on gel formation. Tubes are then inverted in a test-tube stand so that their mouths dip into concentrated solutions, or crystals, of salts of the transition elements. Copper sulphate is a good one, or any of the organic dyes which diffuse much less rapidly than inorganic substances, on account of their larger molecules. An excellent variation is to fill one tube with a gelatine sol containing 5 cm³ universal indicator in 100 cm³ sol so that the gel will appear greenish-blue; this tube is then inverted in a few cubic centimetres of 2×10^3 mol m⁻³ (2.0 M) HCl.

A cardboard scale should be placed behind each test-tube, so that the distance diffused can be read off.

Beware of the speed with which the H⁺ diffuses — it may well reach the top of the test-tube overnight before the practical class is due.

It should be noted that the direction of net-diffusion in this experiment is against gravity.

1.4   Adsorption, liquid on colloid

Brownian movement is a manifestation of the kinetic energy of matter and diffusion is one of its consequences. Adsorption, on the other hand, results in a restriction of kinetic movement. This becomes apparent by a release of energy in the form of heat and by a reduction in the volume of the total system on account of the orderly packing of particles on being adsorbed. Adsorption and consequent energy release, as well as concentration changes, have relevance to enzyme action.

A level teaspoon-full of *dry* starch (1.5 g) is placed in a 50 cm³ beaker held at its rim (to avoid warming it) and gently stirred with a thermometer to measure the temperature of the starch; *thereafter* the temperature of distilled water, also kept in a small beaker, is measured with the same thermometer. After drying the thermometer and while one student continues stirring the starch gently but quickly, the other adds 3 drops of the distilled water from a Pasteur pipette. As soon as the water has been stirred into the starch, the temperature is read quickly, the stirring resumed and the procedure repeated until the temperature of the mixture begins to fall. A rise of 8 °C can be expected. The results should be graphically presented, when it will be seen that the temperature increase is steepest at the beginning when all adsorption sites are free (cf. Fig. x in Part A).

Starch powder must be spread out on paper, dried overnight in an oven at about 70 °C and then put into absolutely dry screw-top jars to cool. If this is not done, the starch will be moist to begin with and the results very disappointing.

1.5   Adsorption, mutually between water and ethyl alcohol

The temperatures of 250 cm³ each of water and ethyl alcohol kept in separate 250 cm³ volumetric flasks are recorded, and then the water is transferred to a 500 cm³ volumetric flask, followed by the alcohol. If properly done, the alcohol will float on top of the water. The combined volumes will be a little less than 500 cm³. The flask is now inverted and

Strictly, the accuracy of the three volumetric flasks should be demonstrated first, with water in the two 250 cm³ flasks combining to fill the 500 cm³ flask to the mark. However, this leaves the flask that is to hold the alcohol wet — definitely not wanted.

All transfers should be done via funnels.

When the alcohol is poured into the flask containing the water, it should not run down the sides. Therefore, the flask should be held vertical and the alcohol should slowly and gently splash on to the surface of the water. Some alcohol will mix at first, but the bulk will form a separate layer. This needs a little practice.

Both volume contraction and temperature elevation take a minute or two to be completed; by proceeding at leisure, maximum changes are recorded.

The same pipette must be used for all liquids and washed when the liquid is changed. The order should be as indicated. The turpentine must be left to the last as it is difficult to wash out.

It may be noted that the instrument used to measure surface tension is the same as that used for measuring viscosity (cf. Expt 2.5), but the methods of use are different. Viscosity is a flow phenomenon, depending on the ease with which molecules slide past each other; surface tension is due to asymmetrically acting intermolecular forces. Water with an exceptionally high surface tension has a low viscosity; most oils with low surface tension have high viscosities.

the contents are quickly mixed once or twice, when it will be seen that the volume has markedly decreased. The temperature of the mixture is now taken. It will have risen to about 8 °C above ambient temperature (this amounts to a release of approximately 16 000 J (4000 cal)). The volume is now made up with water from a graduated 10 cm³ pipette to the mark of the flask – about 15 cm³ or 3% will be required.

1.6 Surface tension

Membranes form at interfaces of colloidal systems where by one mechanism of adsorption substances concentrate that are able to lower the surface tension of the continuous medium – which in biological systems is water. Biological membrane formation has an element of this process occurring at colloidal interfaces existing between cell wall and protoplast and around all organelles. Therefore, students should be made aware of the exceptional surface tension of water, which is also important in capillary phenomena and penetration of air spaces (cf. Expt 8.5 anatomy).

To determine surface tensions of different liquids, a comparative method is recommended. Suitable liquids are water, water plus a small amount of detergent (e.g. resin or a surfactant), alcohol and turpentine oil. Ten drops of the chosen liquids are allowed to fall from a vertically held, graduated 1 cm³ pipette at as uniform a speed as possible. The volume of the 10 drops should be read to the third decimal place and the mean of three determinations is a reliable measure.

When the volumes of the 10 drops for each liquid have been determined, the surface tension is arrived at by the following expression:

$$\text{surface tension} = \frac{10^{-3} \times 73 \times \left(\begin{smallmatrix} \text{sp. gr. of} \\ \text{unknown} \end{smallmatrix}\right)\left(\begin{smallmatrix} \text{volume of} \\ \text{10 drops of} \\ \text{unknown} \end{smallmatrix}\right)}{\text{volume of 10 drops of water}} \text{ N m}^{-1} \quad \substack{(10^3 \text{ dyn cm}^{-1} \\ \text{or } 10^5 \text{ dyn m}^{-1})}$$

When water is used as the 'unknown', this expression equals $10^{-3} \times 73$ N m^{-1} (73 dyn cm^{-1} or 7300 dyn m^{-1}) which is the surface tension of water; hence this method of determining surface tension is comparative and not an absolute measure.

POTENTIAL HYDROGEN ION CONCENTRATION AND COLLOIDAL PROPERTIES

pH control of metabolic processes is common to all living systems and therefore students must have a good understanding of this physicochemical concept. No attempt is made here to deal with the theory of pH, but a few exercises with biological implications are outlined.

1.7   pH dependence of water on dissolved carbon dioxide

Students are asked to blow through a pipette into a 50 cm³ beaker containing 5 cm³ water at pH between 6.5 and 7.0. The pH is indicated by universal indicator and the colour change observed. This has implications for the pH of the water in leaf cell walls, in the dark very different from that in the light.

1.8   pH and silver binding by weak acidic groups
Based on information given by Professor D. H. Jennings, University of Liverpool

The presence of weak acidic radicles on and within the cell walls of yeast can be demonstrated by pH treatments of yeast suspensions (see Part A), 5 cm³ of which are placed into each of six 50 cm³ plastic centrifuge tubes for centrifuging at 2200 g for 5 min. The pellets are then resuspended in 10 cm³ of 25 mol m⁻³ (25 mM) citrate buffer (see Part C1) at pH values of 3.0, 4.0, 5.0, 6.0, 7.0 and 8.0 and left for 40 min, after which time the degree of sedimentation at the different pH treatments is scored.

After spinning down and washing by resuspension in 10 cm³ ice-cold water, the suspensions are once more pelleted, resuspended in 10 cm³ of 10² mol m⁻³ (0.1 M) AgNO₃ and left for 5 min in the dark.

The suspensions must be finally pelleted and washed twice with 5 cm³ ice-cold water before suspending the last pellet in 0.25 cm³ distilled water and transferring the resulting slurry on to three layers of Whatman No. 1 paper. This preparation is kept in the light for 30 min. The resultant degree of brown colouring at the different pH treatments is taken as an indication of the amount of silver associated with the acidic radicles on the yeast cell walls, because of the different degrees of dissociation at the different pH values (cf. Expts 1.9, 1.11, 1.13 & 11.11).

The degree of sedimentation of the yeast depends largely on the charges on the cell walls. Uncharged cells tend to coagulate; like charges repel (cf. Expt 1.13).

1.9   Isoelectric range of a protein, experimental determination

The amphoteric properties of proteins constitute one of their outstanding characteristics and this has a bearing on enzyme action. The isoelectric range in which proteins are insoluble and inactive can be determined by a crude titration procedure (see below) and by calculation (see Expt 1.10).

A solution of 1 g casein in 100 cm³ distilled water at pH 7.6 (adjusted with KOH) is prepared by adding the casein to 60 cm³ boiling water while stirring until it is dissolved. The volume is then made up to 100 cm³. The resulting sol is clear but has a very slight bluish opalescence. After the sol has cooled to room temperature, 2 cm³ are placed in a 50 cm³ beaker and the pH is taken with indicator paper. This is followed by the slow addition of drops of 20 mol m⁻³ (20 mM) HCl, one drop at a time. Where the drop falls, a white precipitate will be seen but, on swirling the liquid round, it will disappear. This is repeated until the white cloudiness does not properly disappear but the whole volume of liquid assumes a denser bluish haze than before. At this point the pH is taken again and it represents the upper limit of the isoelectric range. The

The dropwise addition of acid and alkali is essential. If students cannot control the delivery from pipettes, burettes may solve the problem.

Swirling after each addition of reagent must be thorough, especially near the limits in order to avoid 'overshooting'. Sometimes 'curdling' or precipitation may occur and rather longer mixing and swirling will be required to redissolve the precipitate.

Observation of the sol is best against a dark background, e.g. the bench top or the floor.

addition of acid continues drop by drop with thorough swirling and mixing; the bluish-white haze will first increase markedly but then the area where further drops of acid fall into the sol will become clear again. When, after mixing, the sol is clear, the pH is taken once more, representing the lower limit of the isoelectric range. The whole process is then repeated in the reverse direction with the same sample, using 20 mol m^{-3} (20 mM) KOH, reaching first the lower and finally the upper limit of the range.

Lower and upper limits of the range determined with HCl and with KOH will not be the same because of some hysteresis. Therefore, the mean of the two determinations for each limit must be taken.

1.10 Isoelectric range of a protein, determination by calculation

Based on information given by Dr R. Phillips, University of Stirling

It is critical that NaOH and acetic acid are equimolar. Because NaOH can 'go off' in its bottle, it is strongly recommended that the NaOH solution is titrated to ensure that it is 2 × 10^2 mol m^{-3} (0.2 M).

Five cubic centimetres of a sol of 1 g casein in 100 cm^3 of 2 × 10^2 mol m^{-3} (0.2 M) sodium hydroxide are diluted with 5 cm^3 of 2 × 10^2 mol m^{-3} (0.2 M) acetic acid, giving 10 cm^3 of a 10^2 mol m^{-3} (0.1 M) sol of sodium acetate containing casein. Of this sol, 1 cm^3 is added to each of six test-tubes made up as shown in Table B1.10.

Each test-tube will contain 10 mol m^{-3} (10 mM) sodium acetate, 0.005 g casein and some acetic acid; the contents of all tubes must be

Table B1.10 Quantities of reagents used for the visual determination and scoring for degree of opalescence of casein sols near their isoelectric range.

distilled water (cm^3)	8.75	8.21	6.50	8.21	6.50	1.10
2 × 10^2 mol m^{-3} (0.2 M) acetic acid (cm^3)	—	—	—	0.79	2.50	7.90
20 mol m^{-3} (20 mM) acetic acid (cm^3)	0.25	0.79	2.50	—	—	—
resulting concentration of acid (mol m^{-3} or mM)	0.50	1.58	5.00	15.80	50.00	158.00

thoroughly and quickly mixed before and after the addition of the protein sol. After 15 min the tubes are examined and scored for their degree of opalescence so that rough limits of the isoelectric range can be discerned. The protein is within its isoelectric range where opalescence is greatest.

Solutions containing acetate and acetic acid assume relative proportions of each component according to the equation:

$$\text{equilibrium constant, } K_{EQ} = \frac{[\text{acetate}]\,[\text{H}^+]}{[\text{acetic acid}]}$$

where $K_{EQ} = 2 \times 10^{-5}$.

Because three of the terms in the equation are known to a first approximation for each solution, the fourth, [H$^+$], can be calculated, and pH = $-\log_{10}$ [H$^+$].

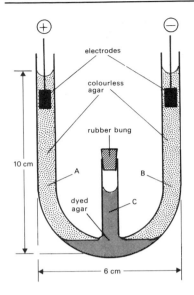

Figure B1.11 Longitudinal section of apparatus used for electrophoretic dye migration.

1.11 Electrophoresis

Proteins migrate differentially in electric fields, depending on the pH of the medium in which they are present. At their isoelectric point this migration is absent. Electrophoresis has become an important analytical technique. As most proteins are colourless, electrophoretic movement must be followed by staining techniques to identify particular proteins. The experiment suggested here uses dyes migrating differentially in an electric field to illustrate the process of electrophoresis.

A U-tube as shown in Figure B1.11 is filled via tube C with a sol of 1.5 g agar dissolved in 100 cm³ water and stained with one of the selected dyes to form a plug as illustrated. After this plug has set into a gel and tube C has been closed with a rubber bung, limbs A and B are filled with plain agar of the same concentration as the dyed sol. Electrodes are implanted and the assembly is left until the agar has gelled. A second assembly with a second dye is set up for comparison. A d.c. current of between 80 and 90 V is then applied and the dyes will move to their respective poles.

Recommended dyes are:

malachite green
toluidine blue   } moving to the negative pole
aqueous methylene blue

methyl red
thymol blue   } moving to the positive pole
bromophenol blue

1.12   pH changes on root surfaces
cf. Expt 11.11

Changes in the pH of a medium surrounding the roots of young maize seedlings can be made visible and the pH values estimated by the colour changes of bromocresol purple indicator incorporated in the growth medium.

Six- to eight-day-old maize seedlings (*Zea mays*) grown between moist papers or in moist sand are needed for the experiments. They are used in an adjusted nutrient medium (see Part C1) with either $Ca(NO_3)_2$ anhydrous or $(NH_4)_2SO_4$ as the nitrogen source. The pH of the medium is adjusted with NaOH to either 4.5, 6.0 or 7.0, depending on the experiment to be executed.

One hundred grammes of a *purified* agar medium, Agar-Agar from Merck Chemicals (see Part C2), Ionagar from Oxoid Ltd (see Part C2) or Agarose from Sigma Chemicals (see Part C2) are mixed with 0.006 g bromocresol purple and the specific nutrient solution to be used, to result in a final agar concentration of 0.75 g 100 cm⁻³. The preparation of this medium is carried out at 40 °C and the roots of the preculture-grown seedlings are placed in the medium at about 30 °C. The medium is contained in a flat, transparent Perspex vessel (12 × 25 × 1.5 cm). It must be ensured that the roots are positioned vertically and when this has been done, the assembly is cooled rapidly to 20 °C to obtain a solidified medium.

With the medium at pH 4.5 (yellow) containing ammonium sulphate

A half-strength nutrient solution similar to the one shown in Part C1 with one-fifth strength micronutrients was used by Marschner *et al.* (1982).

As is usual, root-hair formation is almost absent in liquid or agar growing media.

If the embedded roots require to be held more firmly, an additional thin layer of liquid but cool agar can be applied above the solidified agar medium.

as the nitrogen source, no colour change will occur, because when $(NH_4)_2SO_4$ is taken up protons are released, so that the acidic medium remains acidic. However, in the medium at pH 4.5 containing $Ca(NO_3)_2$, the yellow colour will change to purple around branch roots and behind the root tip, indicating a rise in pH to between 6.0 and 7.0 within 6 h, because when nitrate is taken up OH^- is released into the medium.

In the experiments using the culture medium at pH 6.0 (red), acidification to about pH 4.0 (yellow) will occur within 5 h on account of proton release with $(NH_4)_2SO_4$ as nitrogen source, whereas in the medium with $Ca(NO_3)_2$ no colour change will occur, because the high pH of the medium will remain high when OH^- is released.

REFERENCE

Marschner, H., V. Röhmheld and H. Osenberg-Neuhaus 1982. *Z. Pflanzenphysiol.* **105**, 407–16.

COLLOIDAL PROPERTIES

1.13   Stability of colloidal systems

The properties of colloidal systems, including their exceptional stability, are of particular relevance for all living systems. Most organic colloidal systems have two stability factors: (a) the mutual repulsion of particles carrying like charges (usually negative) and (b) the cushioning effect between particles during their kinetic movement of 'water shells' held by surface sculpture phenomena, electropotential forces or by both. Many inorganic colloidal systems, e.g. clay soils, have only one stability factor, namely that of the electropotential.

(a)   Two 1000 cm³ measuring cylinders filled to near the top with a clay sol in distilled water are placed side by side. To one of these cylinders are added *without stirring* 3 cm³ of 5×10^3 mol m^{-3} (5.0 M) HCl from a pipette. After a comparatively short time when the fast-diffusing H^+ ions have neutralised the negatively charged clay particles, the latter agglomerate and settle out.

(b)   For comparison, 3 cm³ of 3 g *gum arabic* dissolved in 100 cm³ water are put into a test-tube to which a level teaspoon-full of $AlCl_3$ crystals has been added. This will remain a clear sol. However, if 3 cm³ *absolute* alcohol are now added, precipitation will occur as the second stability factor is removed by the dehydrating properties of the alcohol. The experiment should also be carried out in the reverse order.

The alcohol must be truly absolute.

1.14   Cytoplasmic streaming
cf. Expt 3.6

Cytoplasmic streaming is an expression of physiological activity; it involves actin and myosin in microfilaments for the generation of motive force, in addition to localised changes in the colloidal state of cytoplasm (sol–gel), pH and intermolecular forces. To demonstrate it, the most suitable materials are *Chara* and *Nitella* cells and the leaves of *Elodea canadensis* (see Part A). Although irregular, this streaming is an organised movement in very telling contrast to the ceaseless, random Brownian movement – for this reason it is mentioned here and should be observed as a comparison to Brownian movement (Expt 1.1; see also Expts 3.6 & 4.2).

Chara and *Nitella* cells appear to exhibit this phenomenon at almost any time; *E. canadensis* leaves are best observed after they have been kept in the dark overnight and are then taken from near a growing point with good, smooth forceps. After illumination at a little more than critical illumination for a few minutes, during which the leaf should be scanned under 100× magnification, movement will be detected. It is generally stated that cells near the leaf apex and the elongated cells in the middle of the lamina are most reliable in exhibiting cytoplasmic movement. However, the cytoplasm of marginal cells and those further down towards the leaf base is often equally active and does in fact show the circular cyclosis very well. Once movement has been detected, it should be studied under 400× magnification for some minutes, as it will speed up under illumination. When a steady state has been reached, rates of streaming can be measured across a selected range of eyepiece micrometer divisions (see Part A).

Often streaming will only gradually commence under illumination and increase with time. Detection of movement under 100× magnification should be followed by continuous observation under 400× magnification.

For more serious investigations of cytoplasmic streaming in *Elodea*, its dependence on temperature can be studied if suitable microscope stages are available or, using *Chara* or *Nitella*, if streaming in the cells is observed with the material submerged in water in Petri dishes. Reasonable temperature control can then be achieved by using ice-water and water at 25 °C judiciously mixed in the dishes. Determinations of temperature coefficients (Q_{10} values) (see p. 31) over the range 5–15 °C should be possible.

Statistical treatment of all measured values will be necessary for valid conclusions.

REFERENCE

Hope, A. B. and N. A. Walker 1975. *The physiology of giant algal cells*. Cambridge: Cambridge University Press.

RELEVANT ANATOMICAL STUDIES

The morphology of most dicotyledonous shoot growing points is similar to that of *Elodea canadensis* and, since this material is easily dissected by using a pair of forceps and a needle, it is suggested that it be used for this study. The growing point is about 0.05 cm long and becomes recognisable after removal of the leaves from near the tip of a branch kept in a drop of water on a microscope slide. The use of bench lenses and microscopic observation of the dissected growing point are recommended.

2 Enzymes

2.1 Amylase activity and gibberellic acid, (a) Using dialysis

Based on information given by Dr J. Hannay, Imperial College of Science and Technology
cf. Expts 1.3, 2.2, 2.3, 3.1, 10.1, 10.2, 10.10 & 11.5

Seeds should not be soaked before cutting as this may well initiate GA₃ signals to the aleurone layer.

It is best to dehusk the dry seeds by hand – it makes the students familiar with their specific seeds and avoids all possibility of the seeds imbibing water.

To minimise infection by microorganisms, air movement in the laboratory should be minimised and the lid of the Petri dish held over the base while pouring the plate and while placing the half grains with forceps. Colonies of microorganism can produce amylase; this will be apparent if the colonies are separate from the half grains. If infection is on the grain, it is usually localised so that the resulting halo is asymmetric and not concentric with the half grain.

Experimental arrangement (2) could be considered unnecessary if arrangement (4) is successful, because the inside of the membrane containing the embryos will be completely colourless on account of amylase activity but the outside will be coloured unless GA₃ can reach the non-embryonic halves. This occurs in (5) where colourless areas will be in the form of halos around the half grains (see Fig. B2.1b).

A composite experiment can be carried out that tests for the interaction between gibberellic acid (GA_3) and amylase activity in germinating barley (*Hordeum vulgare*) caryopses and demonstrates also the dialysing property of a membrane (cf. Expt 3.1).

Twenty-five *dry* barley seeds without husks are cut so that one half contains the embryo and the other consists of the endosperm and aleurone layer. These half seeds are sterilised for 15 min in a solution of 1 g sodium hypochlorite in 100 cm³ water and then thoroughly washed in sterile distilled water.

The seed halves are used as follows:

1   Five seed halves *without* embryos are placed, cut surface down, in a Petri dish containing a 0.3-cm-deep layer of starch–agar (1 g agar plus 0.5 g *soluble* starch dissolved in 100 cm³ water (see Part C1)).
2   Five seed halves *with* embryos are placed similarly in a second Petri dish equally prepared.
3   In a third Petri dish, a further five *non-embryonic* seed halves are placed on starch–agar containing GA_3 (1 cm³ of a solution of 0.1 g GA_3 dissolved in 100 cm³ water added to 100 cm³ fluid starch–agar).
4   The fourth Petri dish is set up with a dialysing membrane 'boat' (see below and Expt 3.1) in it. Five *embryonic* seed halves are placed inside the dialysing membrane and five *non-embryonic* halves next to the membrane but outside it.
5   Two blank dishes with dialysing membranes can be added, one with GA_3 inside the membrane and *non-embryonic* seed halves outside, the other with amylase inside the membrane and *non-embryonic* seed outside.

The dishes are incubated at 25 °C for 48 h, or until the embryos show a coleoptile and radicle, when pale aqueous iodine solution (see Part C1) is poured over the agar plates to show where the starch has been hydrolysed. This is indicated by a colourless 'halo' around the seed halves.

The results of these treatments show that when the embryo begins to

Clear-cut results can only be expected once the embryos have produced a small coleoptile and radicle. This depends on seed quality and temperature and may require more than 48 h. On the other hand, some seeds produce unduly long roots which lift the seed off the agar. If this happens, it is best to cut off these roots as they develop.

Note that the non-embryonic half seeds do affect starch so that a red–purple colour results when iodine solution is poured on to the agar. There is never a colourless halo.

In industry GA$_3$ is added to malting barley in order to speed up the malting process by increasing the rate of amylase synthesis.

When making the dialysis boat, be sure that the visking tube (or cellophane) is dry, as Sellotape does not stick when wet; autoclave tape may be more suitable.

Ensure that there is no leakage of starch–agar into or out of the boat when preparing the Petri dishes.

germinate amylase is produced, whereas this is not the case in the half grains without embryo. However, if GA$_3$ is present, the non-embryonic halves also produce amylase, as the enzyme is mobilised in the aleurone layer if it receives the hormone signal; under natural conditions GA$_3$ is released by the embryo. The use of the dialysis membrane shows that whereas the hormone signal (GA$_3$) can pass through the membrane, the enzyme cannot.

DIALYSING MEMBRANE 'BOAT'

The dialysing membrane is made from 2.5- to 3.0-cm-wide visking tubing (from Fisons, Part C2) in the form of a 'boat' 6 cm long, 1 cm wide and 1 cm deep. A 10 cm piece of flat, dry visking tubing is slit along one of its edges, unfolded and neatly fashioned around three sides of a wooden template 6 × 1 × 1 cm, the folded ends being held in place by dry Sellotape as shown in Figure B2.1a. Cellophane can be used in place of visking tubing, but it is not as perfect a dialysing membrane. The membrane boat is placed in the fluid agar in the Petri dish and after the agar has set, starch–agar is poured into the boat to a depth of 0.3 cm.

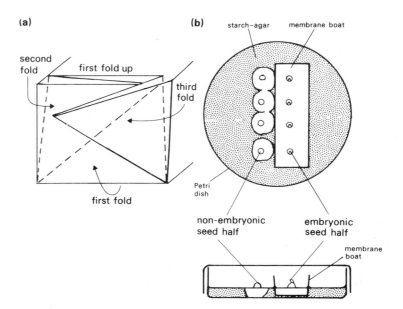

Figure B2.1 (a) Illustration of the sequence of folding a dialysing membrane to fashion a dialysing 'boat'. The membrane is first folded up, then folded from one side following from the other side, each fold overlapping the previous one. Holding the folded ends, these can be secured with Sellotape. (b) Plan and longitudinal section of Petri dish with dialysing 'boat' and germinating embryonic and non-embryonic half grains of cereal.

2.2 Amylase activity and gibberellic acid, (b) Without dialysis

Based on information given by Professor M. A. Hall and Dr A. R. Smith, University College, Aberystwyth
cf. Expts 2.1, 2.3, 10.1, 10.2, 10.10 & 11.5

To remove the husks, the caryopses can be placed for 1 h in a dilution of 50 g H$_2$SO$_4$ in 50 cm^3 water. This is followed by washing in sterile distilled water and drying overnight in a sterile vacuum desiccator.

A variation on Experiment 2.1 can be carried out using non-embryonic half caryopses only, but supplying the starch–agar medium together with 0.001–10 g m^{-3} (10^{-6}–10^{-2} g l^{-1}) GA$_3$ standards. To investigate abscisic acid (ABA) inhibition of GA$_3$, one starch–agar medium can be prepared with (1 g m^{-3} (10^{-3} g l^{-1}) GA$_3$ plus 10^{-2} mol m^{-3} (10 µM) ABA)

The half caryopses should be well spaced on the agar plates so that there is no overlapping of 'clear zones'.

and another medium with (0.1 g m^{-3} (10^{-4} g l^{-1}) GA$_3$ plus 10^{-2} mol m^{-3} (10 μM) ABA). There is no need for the dialysing membrane boat in this variation of the experiment.

The areas of each clear zone are determined for each dish and treatment so that mean areas per treatment can be plotted against log$_{10}$ [GA$_3$]. The results should show that provided there is no ABA present the amount of α-amylase released from the aleurone layer is proportional to the concentration of GA$_3$.

2.3 Amylase activity and temperature
By Professor H. Meidner, University of Stirling
cf. Expts 3.4, 3.5 & 11.5

A tray full of about 100 or more seeds soaks up a surprising amount of water; it is necessary to start with really wet towelling and keep it moist. The plastic envelope round the tray helps a great deal.

Amylase can be obtained from barley (*Hordeum vulgare*) caryopses which have germinated at 20 °C for 72 h between moist paper towels in trays enclosed in plastic bags. After every 24 h it is necessary to check that the papers are still moist.

Amylase is extracted from 10 weighed seeds by grinding the seeds in a mortar to a smooth paste with 1 cm^3 of a stabilising peptone solution (see below). The paste is thinned with another 29 cm^3 peptone solution (total peptone 30 cm^3). After standing in the mortar for about 5 min with gentle stirring, the mixture is centrifuged at 1200 g for 5 min. The supernatant contains the amylase. (If no centrifuge is available, straining through a double layer of cheesecloth may do.)

It is imperative that all students add the correct reagents to their *own* reaction vessels and that the time of addition is recorded. All reaction vessels must be labelled or numbered so that the mixtures can be reliably verified.

It is good practice to run a blank with the peptone alone.

Enzyme extracts must only be added to the reaction mixtures once the treatment temperatures have been established.

Five 50 cm^3 conical flask reaction vessels are needed, each containing 10 cm^3 buffered starch solution (see below) delivered accurately from a pipette, as well as the various amounts of distilled water shown in Table B2.3a for the different temperature treatments, also accurately pipetted. Vessels are placed in their respective water-baths with shaking trays (if no mechanical shaking is available, frequent mixing by hand is essential) at 20, 28, 36, 45 and 56 °C, or any other desired temperatures, to equilibrate for 5–10 min. After equilibration the amount of enzyme extract shown in Table B2.3a is accurately pipetted into the appropriate reaction

Table B2.3 Amounts of buffered starch solution, distilled water and enzyme extract used in each temperature treatment in order to achieve suitably similar times for completion of reactions at different temperatures. All quantities must be accurately pipetted.

(a)	Temperature (°C)				
	20	28	36	45	54
buffered starch sol. (cm^3)	10	10	10	10	10
enzyme extract (cm^3)	9	5	2.5	1.5	1.5
dist. water (cm^3)	1	5	7.5	8.5	8.5
total volume (cm^3)	20	20	20	20	20

(b)	Temperature (°C)			
	20	28	36	45
buffered starch sol. (cm^3)	10	10	10	10
enzyme extract (cm^3)	7	4	2.5	1.5
dist. water (cm^3)	3	6	7.5	8.5
total volume (cm^3)	20	20	20	20

It is impossible to be certain about the amylase activity gained from different seed lots. A trial run must be made before the class starts and, if the reactions are too slow or too fast, the amounts of extracts and distilled water can be adjusted to make the experiment manageable. For instance, if the reaction was too fast, the mixtures could be varied as shown in Table B2.3b.

vessel and the time of addition recorded so as to be able to perform all other tests in rotation.

After 4–5 min (or other time as determined by a trial run), a 0.5 cm³ sample of the reaction mixture from the first vessel is taken out with a bulb pipette and delivered into a test-tube containing 2.5 cm³ iodine solution (see below).

If a blue–black colour results, the starch has not been sufficiently dextrinised. Sample tests must be taken in rotation from all temperature treatments and the cycle repeated after a further 3 min. Sooner or later, a dark, reddish wine colour will appear at one of the temperatures and gradually a rich and finally a golden sherry colour will be obtained with further sampling. Test-tubes with the end-point colour must be prepared beforehand and results of treatments compared with this standard either by eye, comparator or spectrophotometer. A very pale yellow end-point denotes that the reaction has gone too far. This should be avoided.

Results are calculated as follows:

amount of starch dextrinised =

$$\left(\begin{array}{c}\text{amount of starch}\\\text{in 10 cm}^3\text{ solution}\end{array}\right) \times \left(\frac{60\text{ min}}{x\text{ min}}\right) \times \left(\frac{30\text{ cm}^3}{y\text{ cm}^3}\right) \times \left(\frac{1}{z\text{ g}}\right)\text{g h}^{-1}\text{ g}^{-1}\text{ seed}$$

where x is the time in minutes to reach the end-point. y is the volume of extract used in cubic centimetres and z is the weight of 10 seeds in grammes.

For instance, if at 28 °C it took 15 min to reach the end-point with 0.7 g of seeds:

$$\text{rate of starch dextrinised} = 0.2 \times \frac{60}{15} \times \frac{30}{5} \times \frac{1}{0.7}\text{ g h}^{-1}\text{ g}^{-1}\text{ seed}$$

$$= 6.8\text{ g h}^{-1}\text{ g}^{-1}\text{ seed}$$

SOLUTIONS

PEPTONE

Generous quantities of reagents should be prepared. As students have to be given some supply of each reagent there is unavoidable 'waste', especially of the iodine solution.

A solution is made from 20 g peptone dissolved in 1000 cm³ *hot* distilled water; it must be fresh and should not be more than 24 h old.

BUFFERED STARCH SOLUTION

With the exception of the iodine solution, all preparations must be freshly made and should not be older than 24 h.

A buffer is first prepared, using 40 g *anhydrous* sodium acetate and 30 cm³ glacial acetic acid made up to 250 cm³ with distilled water; this has a resulting pH of 4.7. The buffer must be fresh and should not be more than 24 h old (see also Part C1).

Next 20 g *soluble* starch with very little water (no excess at all) are stirred into a fluid paste; this paste is then slowly added to 750 cm³ boiling distilled water while stirring. After boiling for about 4 min, 100 cm³ cold water are added and the solution is allowed to cool a little before adding 50 cm³ of the buffer solution and making up to 1000 cm³ with distilled water. The resultant colloidal sol has a clear appearance with a slightly bluish haze to it. This starch sol, too, must be fresh and should not be more than 24 h old.

DILUTE IODINE

A stock solution is first prepared with 0.5 g iodine crystals and 1 g potassium iodine made up to 250 cm³ with distilled water. The dilute preparation is then made from 4 cm³ of the stock solution plus 40 g potassium iodide made up to 1000 cm³ with distilled water.

2.4 Factors affecting catalase activity
By Professor H. Meidner, University of Stirling
cf. Expts 1.9, 1.10 & 2.3

It would be best to use one large potato for a set of experiments. However, mixtures of several minced potatoes seem to be fairly uniform. The catalase does not seem to deteriorate when the minced potato is left standing for ½–1 h, but all in all freshly minced is preferable and not discoloured.

When cutting off the base of the 5 or 10 cm³ syringe barrel, it is necessary to keep the hacksaw blade cool by sawing slowly, to avoid the plastic melting and the blade getting stuck.

Catalase activity can be measured by the rate of O_2 release from hydrogen peroxide serving as substrate. Catalase is readily obtained from minced raw potato.

Control of pH is by 10 cm³ of a citric acid–sodium phosphate buffer solution at pH values of 4.4, 5.2, 6.5 and 7.5 (see Part C1). This is added to the minced potato in the 50 or 100 cm³ conical flask reaction vessel (A) shown in Figure B2.4.

Enzyme concentration is varied by the amount of minced potato loaded into the reaction vessel from a 5 or 10 cm³ plastic syringe (B). For instance 1, 3 and 5 cm³ minced potato could be used at pH 6.3 and with constant 10 vol. strength hydrogen peroxide (see Part C1) as substrate concentration.

The concentration of the substrate is varied by using 2.5, 5, 10 and 20 vol. strengths of hydrogen peroxide solutions delivered in 5-cm³ quantities from the 5 cm³ syringe (C). All preparations are kept at pH 6.3 with a constant 3 cm³ minced potato as enzyme source.

The rate of release of O_2 is timed with a watch at 30-s intervals, reading off the volumes collected in the burette (D) from the delivery tube (E).

The apparatus is assembled in the following manner:

To obtain the desired level of water in the beaker (F), the beaker is first filled to near the top, and then the inverted burette is inserted to below the water level with a finger kept over the opening. With the burette resting on the bottom of the beaker, the water is decanted until only about 3 cm in depth remain; this reduces the pressure that the escaping O_2 has to overcome.

The reaction vessel must not be held in the warm hand: it must be freely exposed to room temperature and arranged at the correct level relative to the beaker (F) as shown in Figure B2.4.

(1) The burette (D) is filled with water and inverted and placed in beaker (F), which is filled with water.

(2) Minced raw potato is put into the reaction vessel (A) from syringe (B) followed by the addition of 10 cm³ buffer solution at the chosen pH.

(3) The rubber bung, fitted with the delivery tube (E), is moistened with a wet finger to enable an airtight fit to be made without undue pressure, and is fitted into the mouth of the reaction vessel (A). The assembled reaction vessel is positioned in such a way that the mouth of the delivery tube is not underneath the mouth of the inverted water-filled burette (D).

(4) The 5 cm³ plastic syringe (C) filled with hydrogen peroxide of the selected strength is fitted into the rubber bung so that it is airtight and the hydrogen peroxide is injected into the reaction vessel containing the minced potato and buffer solution; air will escape from the delivery tube but must not enter the burette.

Ensure uniform initial swirling of content of reaction vessel (no more than 3 s); this is especially needed when rate of O_2 release is slow. A gentle continuous swirling (though not practicable in a class) would be good, because in all the experiments the rate of oxygen evolution tends to slow down; this is an effect of diffusion of enzyme to substrate limiting the rate of reaction. When plotted, the results will show that both enzyme concentration and substrate concentration appear limiting, which is not possible. What happens is that another factor, namely diffusion of the reactants, limits the reactions and confounds the results.

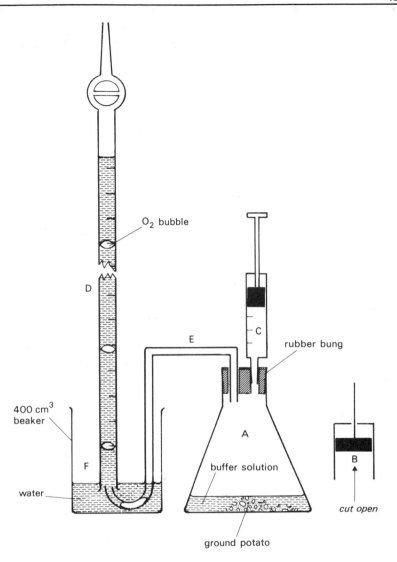

Figure B2.4 Longitudinal section of apparatus assembly for the measurement of catalase activity under various external conditions. The delivery tube is shown in the position for collecting O_2; it must *not* be in that position when the H_2O_2 is injected.

(5) The reaction vessel should now be swirled for 2–3 s and positioned so that the mouth of the delivery tube is beneath the mouth of the inverted burette as shown in Figure B2.4.

(6) The escape of a volume of O_2 is timed every 30 s.

RELEVANT ANATOMICAL STUDIES

PLASMODESMATA AND WOUND CAMBIUM

Potato tissue is used for the experiment on catalase activity for the sake of convenience; most other tissues contain this enzyme. Likewise, the particular anatomical studies are included here for convenience only, as the cells of most living tissues are connected by plasmodesmata and are capable of wound healing – two characteristics of living plant tissue *per se* and very relevant to Experiments 9.1 anatomy, 10.7 and 10.8.

Plasmodesmata

The presence of plasmodesmata connecting living cells within a tissue can be made visible by the staining method of Johansen (1940). Thin hand-sections of potato tuber are good material, but the white endosperm of fresh horse-chestnut seeds is particularly good as it shows pit-fields traversed by many plasmodesmata in areas of narrowed cell walls. The sections need not be large – even pieces of 0.1 × 0.2 cm are good enough; they should be collected in water in a watch-glass and washed in order to allow starch grains from broken cells to settle out. Tapping the sections with a brush or the flat end of the handle of a dissecting needle is good practice. Good sections appear translucent and they will be stained very dark by iodine and crystal violet stains.

The reagents and steps specified by Johansen should be followed closely. Reagents are best kept in watch-glasses and the tissue pieces best handled with camel-hair brushes. While the tissue pieces are immersed in the different solutions, the watch-glass should be agitated by moving it quickly to and fro. Occasional tapping of the tissue pieces with a brush is also recommended to help the release of starch grains from broken cells. When agitated as described above, the starch grains will collect in a line in the centre of the watch-glass at the bottom of the solutions, while the tissue can be moved with a brush to the edge.

The stained and cleared sections should be cleaned with a camel-hair brush and examined in the mounting medium. There will always be areas where the plasmodesmata are especially prominent, often where the cell walls are stained a light violet colour. The many dark dots on the surface of the cells are also plasmodesmata, but seen end-on. One can usually find the best plasmodesmata crossing the cell walls when searching at 100× magnification and then continue for better observation and measurements at 400× and under oil immersion.

Wound cambium

The formation of wound cambium (cf. Expt 9.1 Anatomical studies & Fig. B9.1) can be observed in pieces of potato tissue (2 × 1 × 0.3 cm) kept in a closed box lined with moist tissue paper for about 3 days at 20 °C. Hand-cut sections made across the narrow end of the piece of tissue will show that the margin of the section is formed by three or more rows of narrow, brick-shaped cells arranged in a ladder pattern. If the sections are mounted in pale iodine solution (see Part C1), it will be seen that the cambial cells are practically devoid of starch, whereas the tuber cells are rich in starch.

Table B2.5 Flow-times and table values for the estimation of cellulase activity.

Flow-time (s)	Table value	Flow-time (s)	Table value
45	4.81	24	9.99
44	4.93	23	10.57
43	5.06	22	11.27
42	5.19	21	11.79
41	5.32	20	12.59
40	5.48		
39	5.64	19	13.47
38	5.80	18	14.42
37	5.98	17	15.57
36	6.17	16	16.97
35	6.37	15	18.52
34	6.59	14	20.26
33	6.82	13	22.56
32	7.07	12	25.41
31	7.34	11	28.98
30	7.62	10	33.50
29	7.98	9	39.31
28	8.27	8	47.81
27	8.67	7	60.02
26	9.05	6	78.16
25	9.50	5	111.64

This table has been reproduced by kind permission of Professor N. Lewis, University of California, USA.

REFERENCE

Johansen, D. A. 1940. *Plant microtechnique*, 108–9. New York and London: McGraw-Hill.

2.5   Assay for cellulase (β-1,4-glucan-4-glucanohydrolase)

Based on information given by Dr R. Sexton, University of Stirling
cf. Expts 10.7 & 10.8

Cellulase is involved in auxin-stimulated growth, in abscission, fruit ripening and aerenchyma development. The enzyme can be obtained from the separation layer in abscission zones (Expts 10.7 & 10.8) and from ripe mesocarp tissue of peach, tomato and avocado pear, the last being especially potent. Between 0.5 and 5.0 g tissue are ground up in 5 cm³ salt extraction buffer (see below) at 0 °C in a mortar, allowed to stand for 10 min and squeezed through muslin into a centrifuge tube. The extract is centrifuged at 20 000 **g** for 20 min. The supernatant is carefully taken up with a Pasteur pipette – there must be no debris in the fluid taken up as this would interfere with the assay.

The assay depends on the enzyme cleaving β-1,4-glucan links in a supplied substrate of carboxymethyl cellulose (CMC) (see below). This substance has a very high viscosity, which is reduced as the links are broken and the reduction in viscosity is measured in a simple 'viscometer' (cf. Expt 1.6), consisting of a graduated 0.1 or 0.2 cm³ blow-out pipette. This is mounted vertically in a retort stand and fitted with a 30-cm-long tube at the mouthpiece. The tip of the pipette should be about 10 cm above the bench top.

The liquids must fall from the pipette in drops, and should not be allowed to run down the side of the collecting test-tube.

The flow-time of CMC plus buffer but without enzyme should remain constant at between 25 and 40 s. If it is longer, the substrate must be diluted with more buffer. During the enzyme assay the temperature of the reagent mixture should remain reasonably constant as temperature affects the viscosity of the substrate markedly – normal fluctuations of 5 °C in the laboratory temperature are tolerable.

Except between quick duplicate measurements, the sucking-up of water into the pipette and blowing out should be routine after each measurement.

The first measurement is with water sucked up to above the zero graduation mark. The water is next allowed to drain for exactly 2 s, measured with a stop-watch. The point reached at 2 s must be marked with glass-marking ink – it is the 'water calibration mark'. The second measurement is with 0.8 cm³ CMC at room temperature mixed with 0.4 cm³ extraction buffer in a 5 cm³ test-tube. A vortex mixer must be used for mixing. (No other mixing is quick or effective enough.) This mixture is sucked into the pipette to above the zero mark and allowed to drain. Its flow from the zero graduation mark to the water calibration mark is timed with a stop-watch; it should be about 30 s. After this measurement, and all others, water must be sucked up the pipette and blown out to clear it of all adhering traces of CMC.

The third series of measurements is the enzyme assay: 0.4 cm³ enzyme extract plus 0.8 cm³ CMC are mixed with the vortex mixer and the mixture is sucked up into the pipette to above the zero mark for a measurement of the flow-time from zero to the water calibration mark. This should be repeated quickly in order to get a duplicate reading. The time of day of this reading must be recorded accurately so that the length of time between consecutive readings is known. Repeated measurements are now made during the next 1–2 h until the flow-time has been reduced to between 8 and 15 s. The time of day and the final flow-time between zero and water calibration marks is again recorded, so that the results of the assay can be read off from Table B2.5.

The number of enzyme units can be calculated from the expression:

$$\frac{\text{Cellulase units}}{\text{cm}^{-3}\ \text{extract}} = \frac{\left(\begin{array}{c}\text{table value of}\\\text{final flow-time}\end{array}\right) - \left(\begin{array}{c}\text{table value of}\\\text{initial flow-time}\end{array}\right)}{\left(\begin{array}{c}\text{finishing}\\\text{time of day, h}\end{array}\right) - \left(\begin{array}{c}\text{starting time}\\\text{of day, h}\end{array}\right)}$$

EXTRACTION BUFFER

The extraction buffer has a final concentration of 20 mol m^{-3} (20 mM) Tris–HCl at pH 8.1, 3 mol m^{-3} (3 mM) EDTA and 10^3 mol m^{-3} (1.0 M) NaCl at 4 °C. The salt concentration must be high to remove the cellulase from the cellulosic walls for which it has a high affinity.

Quantities required to prepare 100 cm^3:

Tris–HCl, pH 8.1	0.0121 g
EDTA	0.0336 g
NaCl	5.6 g

CARBOXYMETHYL CELLULOSE SOLUTION

To prepare this solution, 10 g CMC, type 3 H 3 SF, from Hercules Powder Co. (see Part C2) are stirred continuously for 3 days in 1000 cm^3 of 1 mol m^{-3} (1 mM) phosphate buffer at pH 6.1 (see Part C1). If gelatinous lumps form, these must be broken up until a solution with a homogeneous, syrupy consistency is formed. The solution is then quickly autoclaved and can be kept sterile for long periods.

REFERENCE

Almin, K. E., K. E. Erikson and C. Jansson 1967. *Biochem. Biophys. Acta* **139**, 248–53.

2.6 Substrate concentration and xanthine dehydrogenase activity
**Based on information given by Dr R. Phillips, University of Stirling
cf. Expts 1.9 & 1.10**

Xanthine dehydrogenase is an enzyme non-specific in respect of substrate. It is present in *unpasteurised* milk. Methylene blue, acting as hydrogen acceptor, has the advantage of also acting as indicator by its decolorisation. Dilute formaldehyde is a suitable substrate for the enzyme to act upon. During the reaction it is oxidised to formic acid.

Constant amounts of milk and methylene blue (see below) are placed in test-tubes together with variable amounts of buffer solution (see below) as shown in Table B2.6. This mixture is very thoroughly shaken, its temperature is equilibrated in a water-bath and it is shaken once more before the amounts of formaldehyde (see below) shown in Table B2.6 are added and the whole assembly is shaken once again most thoroughly. In

Very thorough mixing is necessary, but it must be gentle so that froth does not form.

Table B2.6 Amounts of reagents to be used for dehydrogenase activity as influenced by substrate concentration. All quantities must be accurately pipetted.

methylene blue (cm^3)	0.50	0.50	0.50	0.50	0.50	0.50
milk (*unpasteurised*) (cm^3)	2.00	2.00	2.00	2.00	2.00	2.00
formaldehyde (cm^3)	0.00	0.15	0.30	0.45	0.60	0.75
buffer, pH 7.0 (cm^3)	2.50	2.35	2.20	2.05	1.90	1.75
total (cm^3)	5.00	5.00	5.00	5.00	5.00	5.00

The somewhat messy liquid paraffin step can be avoided by using disposable syringes instead of test-tubes. When all the mixing and temperature equilibration has been done, the syringe is held with its Luer fitting up and the plunger moved so that no air is left in the syringe barrel. The Luer opening is now sealed with tape, Parafilm or Blu-tack and the syringe submerged in the water-bath.

When liquid paraffin is used, the end-point is the time taken for the mixture to turn white apart from a narrow blue ring at the interface between the paraffin and the mixture.

order to prevent access of atmospheric O_2 to the mixture, a 4-mm-thick layer of liquid paraffin is added, by running it gently down the side of the test-tube held at an angle. The prepared test-tubes are now placed in the water-bath at 25 °C.

The end-point is reached when the mixture has been decolorised. Results can be plotted as 1/time on the y-axis, i.e. rate of reaction, against substrate concentration on the x-axis; or if a straight line is wanted, 1/rate (i.e. time) on the y-axis and 1/substrate concentration on the x-axis.

REAGENTS

METHYLENE BLUE

This is prepared by dissolving 100 mg in 500 cm^3 distilled water.

BUFFER SOLUTION

The solution used is a citric acid–sodium phosphate buffer (see Part C1) at pH 7.0.

FORMALDEHYDE

Commercially available formaldehyde is about 12×10^3 mol m^{-3} (12 M) and a dilution of 12 cm^3 in 500 cm^3 distilled water is suitable for the experiment.

2.7   NAD$^+$ (DPN$^+$)-linked malic enzyme activity
Based on information given by Dr J. M. Palmer, Imperial College of Science and Technology

Plant mitochondria (see Part A) contain an NAD$^+$ (DPN$^+$)-linked malic enzyme capable of producing pyruvate, important in Crassulacean acid metabolism (CAM) and reductive pentose phosphate pathway/4-C dicarboxylic acid pathway (C_3/C_4) metabolism. The presence of this enzyme in detergent-disrupted mitochondria can be demonstrated and its specificity for divalent metal ions investigated.

PREPARATION OF ASSAY PROCEDURE

When malate reduces NAD$^+$, an increase in optical density (D) at 340 nm (3400 Å) occurs. Three milligrammes of protein per cubic centimetre mitochondrial suspension are assayed in 3 cm^3 of a solution of 50 mol m^{-3} (50 mM) MOPS (see Buffers, 'Good', Part C1) buffer at pH 6.8, containing 0.1 cm^3 Triton X-100 in 100 cm^3 water plus 1 mol m^{-3} (1 mM) dithiothreitol.

The assay medium is placed in a cuvette to which is added 0.1 cm^3 of the mitochondrial suspension and after mixing is allowed to stand for 3 min before D is measured in a spectrophotometer at 340 nm (3400 Å). Next, 0.1 cm^3 of 30 mol m^{-3} (30 mM) NAD$^+$ is added, mixed thoroughly with a plastic paddle and the instrument adjusted to zero.

Thorough mixing of the reagents after every addition is essential. Poor mixing is often the cause of inconclusive results because D varies on account of diffusive flow within the reaction vessel.

The rate of reduction of NAD^+ (DPN^+) is rapid and D will increase sharply and then become constant as malate dehydrogenase reaches equilibrium.

MEASUREMENT OF ENZYME ACTIVITY

The first measurement of the rate of changes in D is carried out after adding 50 mm^3 of 10^3 mol m^{-3} (1.0 M) malate at pH 7.0 and stirring thoroughly. Recording of readings of D at 15-s intervals for 2 min can now be started. These will indicate the rate of reduction of NAD^+. This is followed by the addition of 10 mm^3 of 10^3 mol m^{-3} (1.0 M) $MnSO_4$ to the cuvette and readings are taken for another 3 min. When D has reached 0.7–0.8, 5 mm^3 lactate dehydrogenase are added and mixed thoroughly. The changes in D are recorded, showing that pyruvate has been formed and NADH has been oxidised back to NAD^+.

A second experiment can be carried out comparing the rates of enzyme activity in the presence of Mn^{2+}, Mg^{2+}, Na^+ and K^+ using 50 mm^3 of a solution of 10^3 mol m^{-3} (1.0 M) malate, thereby obtaining a comparison between mono- and divalent ions.

RESULTS

The results are tabulated as nmoles NAD^+ reduced per minute per milligramme of protein and interpreted as evidence of the NAD^+-linked activity of malic enzyme in oxidising malate.

2.8 Nitrogenase activity by acetylene reduction
Based on information given by Dr J. M. Wilson, University College, Bangor cf. Expts 4.6 & 4.7

In this experiment use is made of the reduction of experimentally supplied acetylene to ethylene by nitrogenase present in the organism. The chemical change involved is represented by the following equation:

$$HC{\equiv}CH \xrightarrow{\text{nitrogenase}} CH_2{=}CH_2$$

Rhizobium cultures can be obtained from Rothampstead Experimental Station (see Part C2). *Rhizobium* suspensions can also be prepared by crushing a few root nodules from the roots of adult pea or lupin plants; the latter are also suitable material for this experiment.

Suitable material for these measurements are lichens (e.g. *Peltigera canina*) and blue-green algae (*Nostoc*); or *Rhizobium* in leguminous nodules, on which this account is based (see Part A). Good plant material and gas chromatography are essentials. The effects of detaching nodules from the roots and of different temperatures are studied. In order to get some replication, classes are best divided into groups.

PROVIDING EXPERIMENTAL MATERIAL

Four pea seeds are sown in each of several pots; the seeds should be brushed with or soaked in a *Rhizobium* suspension before sowing in N-free sand, although as a rule they will produce nodules without this. To ensure good nodulation, watering with *Rhizobium* suspension at the 2 week stage is recommended. Temperature should never fall below 13 °C. Depending on season, supplementary light may be required.

Successive sowings should ensure that plants 2–3 weeks old (summer) or 3–4 weeks old (winter), and with pink nodules, are available on desired dates. Four- to five-week-old plants are hardly ever suitable. Moderate watering with N-free nutrient solution is recommended and feeding once a week with half-strength, N-containing nutrient solution (see Part C1).

When preparing the material, roots must be gently washed free of sand without damaging the nodules and all excess moisture must be removed with tissue paper, as 'drowned' nodules do not function. Single nodules about 2 mm in diameter and clumps of nodules (do not separate these) can be excised with razor blades, but as the danger of damaging them is considerable, pushing or scraping them off with a sharp fingernail is often better.

Four batches of 20 detached nodules and four comparable portions of suitably trimmed, whole nodulated root systems without shoots are required. The nodules must be pink or white, not green (senescent) and they must be collected on moist filter paper to avoid any drying.

ACETYLENE REDUCTION TECHNIQUE

Each kind of material is placed into a set of four 30 cm³ sample tubes, each containing a 5 × 2 cm moist filter paper strip adhering to one side on to which the nodules can be placed. A fifth tube remains empty, serving as a control.

After sealing the tubes with new Suba-Seal caps (from William Freeman, Part C2), 1 cm³ acetylene is injected via a 1 cm³ disposable syringe into three only in each set of the tubes containing material and into the empty tube in each set. The syringe should be pumped 4 times after each injection to obtain mixing in the tubes and be withdrawn with the plunger pushed right down.

The three tubes in each set containing material *and* acetylene are placed in incubators at different temperatures (e.g. 5, 15 and 30 °C), whereas the tubes without acetylene and the empty tubes serve as controls at any chosen temperature.

After 30 min the first samples of 1 cm³ are withdrawn with new disposable syringes from each of the tubes in turn and injected into the calibrated gas chromatograph. After 60 min the sampling is repeated.

From the recorder output, the amount of ethylene production can be calculated in the usual way and expressed as μmoles per nodule or if nodules are weighed at the end of the experiment as μmoles per gramme fresh weight. Nodules of whole root portions must be carefully detached and all material gently dried with tissue paper before weighing.

The operation of the gas chromatograph and its calibration must be known, and it is recommended to use a 1.5-m-long, 0.04 cm internal diameter glass coil packed with Durapak (from Phase Separations, see Part C2) or similar packing. The oven is operated at 50 °C with nitrogen as the carrier gas, and a hydrogen flame ionisation detector connected to a recorder.

2.9 Phosphoenolpyruvate carboxylase activity
Based on information given by Dr M. Donkin and Dr E. S. Martin, Plymouth Polytechnic

Phosphoenolpyruvate (PEP) carboxylase occurs in many plant tissues. It is involved in carbon fixation into oxaloacetate and malate in 4-C dicarboxylic acid pathway (C_4) and Crassulacean acid metabolism (CAM), it contributes to the maintenance of pH levels, provides intermediates in the Krebs cycle and appears to operate in guard-cell metabolism.

Acetylene is best obtained from a gas cylinder and bled into 500-cm³ conical flasks to make an acetylene–air mixture from which syringes can be filled (cf. Expt 4.7). However, acetylene can be generated from calcium carbide moistened with water and collected in a gas jar (see Part C1).

No naked flames; acetylene is explosive.

The use of new, disposable, 1-cm³ syringes at all times is safest; if this is not possible, very thorough flushing of syringes after each use is essential. Beware of leaking Suba-Seals. We are told that heavy floor wax on laboratory floors can cause strange results — hence the blank control is needed.

Detached nodules are likely to give lower rates of N_2-fixation, being deprived of their energy source from the root system. Generally, the rate of N_2-fixation also declines more rapidly in detached nodules than in whole root systems.

ENZYME EXTRACTION (for solutions, see below)

As a G-25 Sephadex (from Phase Separations, Part C2) glass column (15 × 1 cm) closed with a plug of glass wool in the bottom is required, this must be prepared first by filling it to within 4 cm from the top with pre-swollen G-25. Allow it to settle, pass four column volumes of *extraction* buffer through the column and calibrate with blue dextran in order to find the void volume.

A source of PEP carboxylase is provided by 0.4 g fresh wt *Commelina communis* epidermal strips (see Part A). These are placed in a pre-cooled mortar and ground with a little acid-washed sand in 2.5 cm³ ice-cold *extraction* buffer for 1 min so that no intact tissue remains. The homogenate is poured through one layer of Mira cloth (from Calbiochem, see Part C2) into a graduated tube kept in ice. The mortar should be washed with an additional 0.5 cm³ *extraction* buffer and the volume of extract made up to 3.0 cm³ with some more buffer.

The extract is now passed through the calibrated G-25 Sephadex column, and the void volume as well as several 1 cm³ fractions are collected. Those fractions which show maximum protein absorbance at 280 nm (2800 Å), or which are most turbid, are combined to give about 3 cm³ enzyme extract for the measurements of enzyme activity.

Other tissues could be used and may require a less sensitive setting of the spectrophotometer, for instance: pea pod, pea testa, pea cotyledon, Crassulacean acid metabolism (CAM) plant tissue and 4-C dicarboxylic acid metabolism (C_4) tissue.

ENZYME ASSAY (for solutions, see below)

The assay depends on the utilisation of NADH (DPNH) in a coupled reaction:

As NADH is oxidised, the absorbance at 340 nm (3400 Å) decreases. Assay mixtures are assembled in 1-cm³ cuvettes kept at 30 °C in the quantities and in the order shown in Table B2.9. The double-beam spectrophotometer is set at 0.05 full-scale deflection. Absorbance is recorded at 340 nm (3400 Å) for 5 min after the addition of PEP. After the first 40 s a fairly linear rate of decline should occur.

The enzyme extract and the 8 units of MDH should only be added when the 2 min incubation period can be begun. The reaction is started by the addition of the specific amount of PEP or distilled water to the blank cuvette. The rates may be calculated as μmoles NADH reduced min^{-1} 100 cm^{-3} extract using the molar extinction coefficient, E_M, of NADH, which is equal to 6.22×10^3. The results may then be plotted as:

(a) rate against substrate concentration (mol m^{-3}) (mM); and
(b) 1/rate against 1/substrate concentration (Lineweaver–Burk plot); or
(c) substrate concentration/rate against substrate concentration (Hanes

Table B2.9 Amounts and order of addition of reagents into the 1 cm³ spectrophotometer cuvettes kept at 30 °C. Cubic millimetre (μl) pipettes or syringes should be used.

		Cuvette number					
		1	2	3	4	5	6
	assay buffer (cm³)	0.500	0.500	0.500	0.500	0.500	0.500
	NADH (DPNH) (cm³)	0.1	0.1	0.1	0.1	0.1	0.1
incubation: add only when 2 min. period can begin	10^3 mol m⁻³ (1.0 M) (HCO₃)⁻ (cm³)	0.006	0.006	0.006	0.006	0.006	0.006
	MDH 8 units (cm³)	0.004	0.004	0.004	0.004	0.004	0.004
	enzyme extract (cm³)	0.200	0.200	0.200	0.200	0.200	0.200
add to start reaction*	40 mol m⁻³ (40 mM) PEP (cm³)	0.050	0.020	0.010			
	4 mol m⁻³ (4 mM) PEP (cm³)				0.050	0.035	0.020
	distilled water (cm³)	0.140	0.170	0.180	0.140	0.155	0.170
total (cm³)		1.000	1.000	1.000	1.000	1.000	1.000

* Equivalent amounts of distilled water are used for the blank.

The Michaelis constant (K_m) and maximum rate of reaction (V_{max}) of the enzyme may then be calculated.

SOLUTIONS

Extraction buffer
   50 mol m⁻³ Tris, pH 8.0 (50 mM)
   1 mol m⁻³ EDTA (1 mM)
   5 mol m⁻³ MgCl₂ (5 mM)
   5 mol m⁻³ (5 mM) dithiothreitol, added just before use in order to prevent oxidation of the SH-groups of the enzymes

Assay buffer
   100 mol m⁻³ Tris buffer, pH 7.5 (100 mM)
   12 mol m⁻³ MgCl₂ (12 mM)

NADH (DPNH)
   1 mg cm⁻³ made up just before use

Malate dehydrogenase (MDH; pigeon breast muscle in ammonium sulphate)
   approximately 1000 units in 0.5 cm³ (from Sigma Chemicals, see Part C2)

Sodium bicarbonate
   10^3 mol m⁻³ solution (1.0 M)

Phosphoenolpyruvate (PEP), tricyclohexylammonium salt (from Sigma Chemicals, see Part C2)
   40 mol m⁻³ and 4 mol m⁻³ solutions (40 mM and 4 mM) (from Sigma Chemicals, see Part C2)

G-25 Sephadex
    G-25-100, swollen in extraction buffer (from Sigma Chemicals, see Part
    C2)

Blue Dextran solution
    from Pharmacia Fine Chemicals (see Part C2)

2.10   Non-specific acid phosphatase activity
Based on information given by Dr C. Willmer, University of Stirling
cf. Expts 11.2 and 11.3

Samples of epidermal strips (see Part A) from the lower leaf surface of
Commelina communis are first fixed in acetone for 20 min. After washing the tissue in distilled water for another 20 min, it is incubated for 2 h
at 37 °C in a medium containing 6 cm^3 of 2×10^2 mol m^{-3} (0.2 M) acetate
buffer (see Part C1) at pH 5.4, 2 cm^3 of 50 mol m^{-3} (50 mM)
β-glycerophosphate (disodium salt) and 2 cm^3 of 10^2 mol m^{-3} (0.1 M)
lead nitrate. After incubation the tissue must be washed in distilled water
for 20 min and finally immersed in a solution of 0.5 cm^3 ammonium
sulphide in 100 cm^3 water for 2–3 min. After briefly rinsing in distilled
water and mounting on a microscope slide with coverslip, black lead
sulphide deposits will be observable under $100\times$ magnification, indicating regions of acid phosphatase activity.

Acid phosphatase activity occurs in
vacuoles and spherosomes and other
organelles, but in this test with epidermal tissue the black deposits will be
seen throughout the guard cells, probably due to fixation in acetone, which
breaks down organelle membranes,
releasing the phosphatase.

To verify that the lead sulphide has been enzymatically produced, tests
should be carried out:

(a)   without the β-glycerophosphate;
(b)   by incubation in the presence of 10 mol m^{-3} (10 mM) sodium
        fluoride (1 cm^3 of 10^2 mol m^{-3} (0.1 M) sodium fluoride plus 5 cm^3
        acetate buffer instead of the 6 cm^3 buffer); and
(c)   by using epidermal strips that have been killed by placing in boiling
        water.

Epidermal tissue taken from plants kept in the dark with closed
stomata will have high phosphatase activity in the guard cells, whereas
tissue taken from plants illuminated for 45 min under a plastic hood
(open stomata) will show considerably less enzyme activity as indicated
by the black deposits. All cell walls tend to become blackened.

REFERENCES

Avers, C. J. 1961. *Am. J. Bot.* **48**, 137–43.
Fujino, M. 1967. *Science bulletin of the Faculty of Education, Nagasaki University*,
    no. 18, 1–47.

3 Membranes

3.1 Dialysis
By Professor H. Meidner, University of Stirling
cf. Expt 2.1

In most experiments that involve the passage of substances through membranes, naturally occurring membranes such as plasmalemma, tonoplast or any of those which surround organelles, provide the experimental material. The most important property of these membranes is their *differential permeability*. This varies for different substances. In this experiment (and in Expt 2.1) an artificial membrane is used; this too is differentially permeable.

Bags of visking tube obtainable from Fisons Ltd (see Part C2) are made by thoroughly wetting the flat tube, separating the sides by rubbing, and knotting one end. Then, holding the other end open, the bag can be filled to 3 cm from the top with a weak, *soluble* starch sol (see Part C1). The end is then gripped in a bulldog clip and the filled bag suspended in a 250 cm³ measuring cylinder. Starch sol is then added to the measuring cylinder until near the top. A similar bag is prepared with a pale yellow, aqueous iodine–potassium iodide solution (see Part C1), suspended in another measuring cylinder filled with iodine solution.

When the demonstration is to begin, the bags are removed from their cylinders and are washed thoroughly on the outside under a running tap.

The iodine bag is then suspended in the starch sol cylinder and the starch sol bag in the iodine cylinder. Results can be observed within minutes, the starch sol in the cylinder becoming coloured on the outside of the iodine bag and the starch sol in the bag becoming coloured on the inside of the bag. After a few minutes the colouring is deep and, on lifting the bags out of their cylinders, the effects are very clear.

As a corollary, microscopic sections of potato tuber can be mounted in water and observed while drawing weak iodine solution (see Part C1) across the preparation.

To avoid spilling the bags by inadvertently squeezing them, they should be handled by their clips only.

3.2   Plasmolysis–deplasmolysis
Based on information given by Dr J. M. Wilson, University College, Bangor, and Professor H. Meidner, University of Stirling
cf. Expts 3.3, 5.3, 5.4b & 5.5a

Experiments involving plasmolysis are commonly referred to by that term alone. This is undesirable – the experimental occurrence of plasmolysis should always be coupled with deplasmolysis because only by this means can it be ascertained whether the cells have remained undamaged.

It is also important to point out to students that for cells of land plants plasmolysis is almost exclusively an experimental phenomenon. It can occur only in cells bathed in solutions stronger than their vacuolar sap. Under natural conditions cells of land plants exposed to water stress cannot plasmolyse as there is no solution to fill the space between cell wall and plasmalemma – and air cannot penetrate the wall.

A 10^2 mol m^{-3} (0.1 M) solution of KCl will have nearly twice the solute potential of a 10^2 mol m^{-3} (0.1 M) mannitol solution, because in weak solutions the percentage ionisation of KCl is very high. For comparisons between mannitol or sucrose and KCl, it is the ψ_s of the solutions, not their molarities, that must be considered.

Epidermal tissue (see Part A) of leaves with cells containing anthocyanin sap are most convenient to use. Examples are *Rhoeo discolor*, *Saxifraga tomentosa* and *Cyclamen persicum*. Sections of beetroot tissue, if thin enough, are also suitable. Epidermal strips of *Commelina communis* and *Tradescantia virginiana* are good material after staining with neutral red (0.1 g in 1000 cm^3 water) for 2 min.

For estimates of solute potential (ψ_s) (cf. Expt 5.5a) of cell sap, the plasmolytic and deplasmolytic treatments must be completed quickly to reduce solute leakage, and a non-permeating solute must be used. It is good enough to allow 3 min for the tissue in a solution to estimate the percentage of cells plasmolysed and another 3 min in the next lower concentration for the cells to deplasmolyse. By narrowing the steps of concentration differences, fairly fine estimates of ψ_s can be obtained. The 50% plasmolysis method (Expt 5.5a) is recommended. It must be clearly understood that values of ψ_s determined by this method are for cell sap at incipient plasmolysis, i.e. zero turgor.

Observations are best made at 100× magnification and without a coverslip; only for counting would a higher magnification (and a coverslip) be of advantage. For epidermal tissue, sucrose or mannitol in steps of 2×10^2 mol m^{-3} (0.2 M) up to 8×10^2 mol m^{-3} (0.8 M) are needed. For beetroot tissue solutions, up to 1.5×10^3 mol m^{-3} (1.5 M) may be needed, but a start can be made with 6×10^2 mol m^{-3} (0.6 M). A preliminary test with the chosen tissue must be made in order to select the correct solutions. (For ψ_s of sucrose solutions, see Part C1.)

As a subsidiary experiment, KNO$_3$ solutions in steps of 10^2 mol m^{-3} (0.1 M) up to 4×10^2 mol m^{-3} (0.4 M) for epidermal tissue would show that cells which became plasmolysed in a solution with a ψ_s similar to the mannitol or sucrose solutions used (by no means of the same molarity) would in time deplasmolyse of their own accord in that solution because K$^+$ and NO$_3^-$ can diffuse into the cells (cf. Expt 3.4), whereas sucrose or mannitol cannot.

3.3   Differential permeability
Based on information given by Dr J. M. Wilson, University College, Bangor
cf. Expt 3.2

With substances to which the plasmalemma is not permeable (e.g. sucrose and mannitol) or only weakly permeable (e.g. glucose and some salts), persistent plasmolysis will occur at solute potentials of the external solution lower than those of the vacuolar sap.

Temporary plasmolysis will occur initially with substances to which the plasmalemma is permeable (e.g. urea and ethylene glycol), but deplasmolysis will follow as the solutes diffusing into the vacuole accumulate, and thus raise the solute potential of the sap.

Once the plasmolytic and deplasmolytic process has been observed (Expt 3.2), it can be shown that membranes have different permeabilities for different solutes. Thus a solute to which the membrane is highly permeable may not plasmolyse the cell at all or, if it does, may gradually deplasmolyse it again.

The material recommended for this experiment is epidermal tissue of *Rhoeo discolor*. However, epidermal strips (see Part A) of *Saxifraga tomentosa* and *Cyclamen persicum*, as well as thin sections of beetroot, are also suitable. For each kind of material appropriate ranges of concentrations have to be experimentally determined as solute potentials of cell sap differ with time of year, growing conditions and variety.

A study can also be made of the effects of different substances on membrane permeability. The solutions recommended are: chloroform-saturated water, 10^2 mol m^{-3} (0.1M) HCl, 10^2 mol m^{-3} (0.1 M) NaOH and 25 cm^3 ethanol diluted with 75 cm^3 water. These solutions will not cause plasmolysis by themselves but affect the permeability of the membrane, and treatments should be for different lengths of time (1–15 min) *before* the tissues are exposed to the plasmolysing solutions. The absence or the speed of occurrence of plasmolysis will indicate whether membrane permeability has been impaired.

It is good practice to use the same beetroot for one set of treatments.

It is important that the water added to the samples is at the correct temperature right from the start of each temperature treatment and to ensure that the tissue in the seventh tube has been frozen solid.

All test-tubes must be labelled to avoid errors.

The material should be treated for equal lengths of time, e.g. 3 min, in the following solutions:

sucrose in six steps of 2.5×10^2 mol m^{-3} (0.25 M) to a final concentration of 1.5×10^3 mol m^{-3} (1.5 M)

glucose in four steps of 2.5×10^2 mol m^{-3} (0.25 M) to a final concentration of 10^3 mol m^{-3} (1.0 M)

urea at two concentrations, 5×10^2 and 10^3 mol m^{-3} (0.5 M and 1.0 M)

ethylene glycol at 5×10^2 mol m^{-3} (0.5 M)

glycerol at 5×10^2 mol m^{-3} (0.5 M)

NaCl at 2×10^3 mol m^{-3} (2.0 M)

3.4   Temperature and membrane permeability
By Professor H. Meidner, University of Stirling
cf. Expts 2.3 & 3.5

Cores of beetroot tissue are cut with No. 8 (1.2 cm internal diameter) cork-borers and 28 discs 0.05 cm thick are prepared with a sharp blade. The discs must be washed for 5 min in running tap water and four are then placed into each of seven test-tubes. Into five of these are added 10 cm^3 tap water taken from containers that have been temperature-conditioned in the respective water-baths for each temperature treatment. The water-baths could be at room temperature, 25, 35, 45 and 55 °C, and are preferably fitted with a shaker. Water from a container kept in a refrigerator at about 1–5 °C is used for the sixth tube and the seventh tube is put into a deep freeze or the freezing compartment of the refrigerator without any water added. Temperature treatments should last for about 45 min. After that time the contents of the tubes are gently mixed by inverting, decanted and either arranged by eye in order of degree of colouring or measured spectrophotometrically at 540 nm (5400 Å). The tube with the frozen tissue is allowed to thaw after the addition of 10 cm^3 water at room temperature, mixed, decanted and measured for absorbance.

Results of absorbance *versus* temperature are presented as a graph. Permeability will be seen to be high after freezing (owing to rupture) and at 55 °C (owing to partial denaturing of proteins in the membrane) and fairly stable between 5 and 25 °C.

3.5   Effect of solutes on the response of membranes to temperature
Based on information given by Dr J. M. Wilson, University College, Bangor
cf. Expt 2.3

The preparation of beetroot discs is the same as for Experiment 3.4. Twenty-four test-tubes are required for six temperature treatments and four solute treatments. Each test-tube requires four discs, so that a total of 96 discs are needed. The solute treatments are as follows:

    10 cm^3 tap water
    9 cm^3 tap water + 1 cm^3 decenylsuccinic acid
    9 cm^3 tap water + 1 cm^3 10^3 mol m^{-3} (1.0 M) CaCl$_2$
    9 cm^3 tap water + 1 cm^3 10^3 mol m^{-3} (1.0 M) sucrose

The temperature treatments must be timed accurately.

A concentration of 10^3 mol m^{-3} (1.0 M) $CaCl_2$ may be found to be strong enough to discolour betacyanin which has leaked from the tissue. If this happens, it is recommended to decrease the pH of the medium with a drop of 0.5×10^3 mol m^{-3} (0.5 M) HCl to restore the red colour – an instructive occurrence in a class practical.

Decenylsuccinic acid increases membrane permeability as it can act as a metabolic inhibitor, whereas $CaCl_2$ tends to increase membrane stability as does sucrose, although the latter may have no effect.

Two hundred cubic centimetres of tap water are heated to 70 °C and 16 discs of beetroot tissue treated at this temperature for 1 min; thereafter, four discs are placed in each of the four solute treatments for 1 h with occasional gentle mixing.

By diluting the 70 °C water with cold water, the temperature is reduced stepwise to 60, 50, 40 and 30 °C. Sixteen discs are kept for 1 min at each temperature, before four of the discs are placed into each of the four solute treatments for 1 h. The last 16 discs are kept at room temperature, four in each of the four solute treatments, serving as controls.

At the end of the 1 h treatments, each coloured solution is mixed by inverting the tube and decanted for spectrophotometric analysis at 540 nm (5400 Å). The effects of the four solutes on the response to temperature of the membranes can be represented by graphs for interpretation.

3.6) Hydraulic conductivity (L_p) of plant cells
Based on information given by Professor J. Barber, Imperial College of Science and Technology
cf. Expt 3.7

Hydraulic conductivity or osmotic permeability here includes that of the composite membrane system of the mucilage-covered cell wall, plasmalemma, cytoplasm and tonoplast of a single *Chara* or *Nitella* cell. These cells are large enough to be handled in the experimental arrangement shown in Figure B3.6. The bigger the cells of the species available, the easier it will be to set up the experiment.

The process utilised is termed 'trans-cellular' osmosis and its rate is measured by the movement of a meniscus in a capillary tube.

Cell water potential (ψ_{cell}) can be assumed to be zero for the cell in water. This will be reduced to the solute potential (ψ_s) of the solution added to compartment 1. Therefore the difference in ψ_{cell} is equal to ψ_s of the added solution.

$$\text{Osmotic permeability, } L_p = \frac{\text{rate of flow (cm}^3\text{ s}^{-1})}{\text{area of cell surface (cm}^2) \times \text{difference in } \psi_{cell} \text{ (MPa)}} \text{ cm s}^{-1}\text{ MPa}^{-1}$$

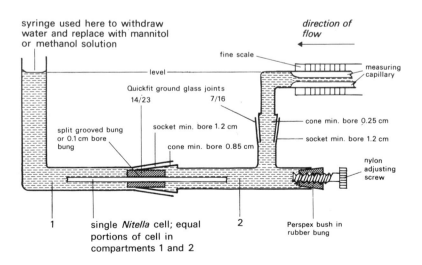

syringe used here to withdraw water and replace with mannitol or methanol solution

direction of flow

fine scale

level

Quickfit ground glass joints
14/23   7/16

measuring capillary

cone min. bore 0.25 cm

socket min. bore 1.2 cm

split grooved bung or 0.1 cm bore bung

socket min. bore 1.2 cm

cone min. bore 0.85 cm

nylon adjusting screw

1

single *Nitella* cell; equal portions of cell in compartments 1 and 2

2

Perspex bush in rubber bung

Figure B3.6 Longitudinal section of apparatus used for the measurement of transcellular osmosis. Special care must be taken to ensure that water levels in compartments 1 and 2 are the same.

After setting up the assembly as described below, both compartments and the capillary tube are filled with tap water without allowing air bubbles to be trapped. The assembly should be kept in a water-bath at constant temperature. The meniscus can be set with the plastic adjusting screw and when it has become steady, equilibrium will have been reached. The tap water can now be withdrawn from the open compartment with a syringe and replaced with one of three strengths of sucrose solution (see Part C1), e.g. 50 mol m^{-3}, 10^2 mol m^{-3} and 2×10^2 mol m^{-3} (50, 100 and 200 mM). Osmotic flow into the sucrose solution will begin almost immediately and meniscus readings should be taken every 10, 15 or 20 s. The movement will be very small and, within 1 min, depending on the molarity of the sucrose solution, 0.05–0.5 cm may be all that will be recorded. It is only the movement during the first minute that is used in the calculation, but the measurement should be made for 3–5 min. This will show the gradual decline in the rate of osmotic flow, because solute movement within the cell, and indeed in its immediate surroundings, will be affected and alter osmotic gradients. The cell also decreases in volume and effects on the composite membrane system may occur.

For the calculation, the volume of the capillary tube in cubic centimetres should be known (fill with mercury for a measured length, weigh at 20 °C and multiply by 13.6). The surface of the cell in each compartment must also be estimated separately (length of cell portion $\times 2\pi r$ cm^2).

The measured rates of meniscus movement in the different sucrose solutions can now be converted to volume rates of flow and the hydraulic conductivity can be calculated as follows:

$$L_p = \frac{\text{sum of surface areas of cell portions in each compartment}}{\text{product of surface areas of cell portions in each compartment}} \times \frac{\text{volume rate of flow}}{\psi_s \text{ of sucrose solution}} \quad \text{cm s}^{-1} \text{ MPa}^{-1}$$

ASSEMBLY OF APPARATUS

Fitting the cell in a watertight manner into the apparatus is the most skilled operation required. There are two methods:

(a)   The use of a very lightly greased, split rubber bung with a suitable narrow and shallow groove, but otherwise with smooth matching surfaces. This, with the cell enclosed, is fitted into the opening of the glass compartment as shown in Figure B3.6.

(b)   The use of a rubber bung tapering by 0.1–0.2 cm in diameter and with a hole of less than 0.1 cm diameter into which the cell can be threaded; the tapered bung with the cell in its bore is fitted into the open end of one of the glass compartments and as it is gently pushed further in, the bore is compressed slightly, making a watertight seal.

The second method is the better of the two, as grease is an unreliable seal but, if grease is used, it should be of the 'rubber grease' or vacuum grease variety. Possibly a combination of both methods may be successful.

The cell should be examined before and after the experiment to ascertain whether normal cytoplasmic streaming is occurring (see Expt 1.14).

At either end of the cell some tissue of the next cell should be left for handling with forceps.

For fitting into the split rubber bung, the cell should be gently dried with paper tissue in the region where it goes through the bung. About an equal length of cell should be positioned in each compartment.

The capillary tube should be cleaned with chromic acid so that the meniscus does not stick and moves smoothly. A good scale, which may require the use of a magnifying glass, must be fitted to the capillary.

The apparatus is otherwise readily assembled either with ground glass joints as indicated in Figure B3.6 or with good fitting plastic tubing, in which case the assembly once set up must not be moved as this will alter its volume and move the meniscus. Glass joints must be greased.

3.7   Measurement of reflection coefficients

Based on information given by Professor J. Barber, Imperial College of Science and Technology
cf. Expt 3.6

In Experiment 3.6 a non-permeating solute (sucrose) was used. If permeating solutes are involved, as they occur in nature, these affect the movement of both solute and water. The permeability that was measured in Experiment 3.6 is valid only for water movement in the absence of a permeating solute; in its presence the combined effects on permeability can be accounted for by the 'reflection coefficient' valid for such a solute (cf. Meidner & Sheriff 1976).

The ψ_s of both solutions must be the same for the calculation of the reflection coefficient. For non-permeating solutes the coefficient equals 1.0, for completely permeating solutes it is zero, but for most solutes it will be between zero and 1.0. It is important to consult Part C1 for ψ_s of sucrose solutions.

In an experimental arrangement of the same kind as in Experiment 3.6, sucrose solutions are replaced with methanol solutions of the same ψ_s (see Part C1). Methanol is a readily permeating solute. By measuring the volume rates of flow in the presence of methanol, the reflection coefficient (δ) can be determined:

$$\delta = \frac{\text{volume rate of flow with methanol}}{\text{volume rate of flow with sucrose}}$$

REFERENCE

Meidner, H. and D. W. Sheriff 1976. *Water and plants*, 103. London: Blackie.

4 Mineral nutrition

4.1 Electropotential measurements and ion uptake
**Based on information given by Dr D. J. F. Bowling, University of Aberdeen
cf. Expts 4.2 & 4.13**

The electropotential difference across exuding plant roots can be measured with silver–silver chloride or calomel electrodes, one dipping into the solution bathing the roots, the other into exuded sap collected in a piece of rubber tubing fitted over the stump of a de-topped plant (cf. Fig. B4.1 & Expt 5.3). Tomato (*Lycopersicon esculentum*), cotton (*Gossypium* spp.), castor bean (*Ricinus communis*) and sunflower (*Helianthus*

It is advisable to set up four plants per culture vessel and support the plants by polythene foam or cotton wool in the neck of the vessel.

The aerating system of the nutrient solution may need to be turned off during measurements of potentials.

Ready-made reference electrodes of the type used for pH measurements are preferable to home-made ones, and must be mounted in retort stands and clamps, not held by hand.

Zero error must be determined first by placing both electrodes in a 1 mol m^{-3} (1 mM) KCl solution. The negative electrode is to be used in the exudate.

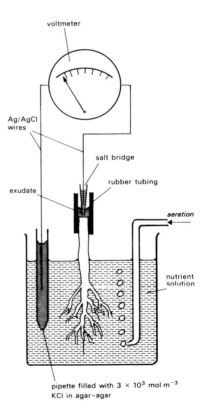

Figure B4.1 Longitudinal section of apparatus used for electropotential measurements across exuding root systems.

Dilute nutrient solutions (see Part C1) give more negative potentials than concentrated solutions.

For monovalent anions the expression for the Nernst potential has its quotient inverted:

$$E_{Cl^-} = 58 \log_{10} \frac{[Cl^-] \text{ exudate}}{[Cl^-] \text{ medium}} \text{ mV}$$

and the expression for H^+ is:

$$E_{H^+} = 58 \text{ (pH exudate} - \text{pH medium) mV}$$

Since pH values are readily measured for both sap and medium, these measurements should be included in this exercise.

annuus) are suitable species. The plants should be growing in aerated nutrient solutions (see Part C1) and have stem diameters of about 0.4 cm.

The electrodes are connected to a millivoltmeter (a pH meter with expanded millivolt scale may be sensitive enough). Measurements of between -50 and -100 mV can be expected. If the apparatus remains connected for 36 h and the output is recorded, diurnal rhythms may be recognised.

By analysing samples of exuded sap and of the nutrient solution for potassium with a flame photometer, the Nernst potential for K^+ ($E_{K^+} = 58 \log_{10}$ {[K^+] medium/[K^+] exudate}) can be calculated and compared with the measured negative potential of the sap in order to determine whether K^+ uptake was 'active' or passive. If the Nernst potential is more negative than the measured potential difference, this suggests active uptake. If both potentials are similar, this suggests passive uptake. The use of 10^{-1} mol m^{-3} (100 μM) potassium cyanide (KCN) or dinitrophenol (DNP) as metabolic inhibitors in the nutrient solution may reduce the measured potentials.

4.2   Ionic relations of Characean cells
Based on information given by Dr J. Collins, University of Liverpool
cf. Expt 4.1

It is advisable to acclimatise the plants by placing them in artificial pond water overnight before the class starts in the laboratory.

If no pond mud is available, sterilised garden soil can be substituted.

Cells should never be allowed to dry out when preparing them for the release of the vacuolar sap and they must not be squeezed in any way as this would release the cytoplasm; only the vacuolar sap is wanted.

If used for observing cytoplasmic streaming, cells should be kept immersed in water in Petri dishes and observed at 100× or 200× magnification (cf. Expt 1.14).

Ecorticated species such as *Chara corallina*, *Nitella flexilis* and *Nitella obtusa* are suitable material for determinations of ionic concentrations in cell sap. The material is best collected fresh from nearby ponds and kept in a large chromatograph tank in pond water or artificial pond water (see Part C1) at 15 °C, not exposed to full sunlight. The plants can grow roots in such a tank containing a layer of sterilised pond mud about 5 cm deep.

Fully turgid cells of up to 8 cm in length with a turgor pressure potential (ψ_P) of about 0.8 MPa are held vertically above a slide coated with solidified wax and are cut cleanly at their lower end with a new sharp blade. The sap will spurt out and can be taken up in 5-mm^3 capillaries from the droplet of released clear, non-viscous cell sap; this is diluted with distilled water to 5 cm^3.

Concentrations of K^+ and Na^+ can be determined by flame photometry, K^+ also with Merckoquant test sticks (BDH, see Part C2; see Expt 9.2) and Cl^- by titration. Generally, K^+ is present in about 78 mol m^{-3} (78 mM) strength, Na^+ in about 60 mol m^{-3} (60 mM) and Cl^- in about 150 mol m^{-3} (150 mM) strength, whereas in pond water the concentrations are approximately 10^{-1}, 1.0 and 1.3 mol m^{-3} (100 μM, 1 mM and 1.3 mM), respectively.

If electropotential measurements (see Expt 4.1) are not made, the following approximate values for potentials taken from the literature will allow for calculations using the Nernst equation:

$$\begin{array}{c} \text{electropotential} \\ \text{difference between} \\ \text{vacuole and medium} \\ \text{at 20 °C} \end{array} = \frac{58}{\text{valency}} \log_{10} \left(\frac{\text{outside molarity}}{\text{vacuolar molarity}} \right) \text{mV}$$

Thus, for Cl^- the Nernst potential would be

$$\frac{58}{1} \log_{10} \frac{1.3}{181} \text{ mV}$$
$$= +119.8 \text{ mV}$$

whereas a measured value of about -160 mV is commonly found; it should be noted that the potential sign is given with reference to the vacuole.

For K^+ the Nernst potential will be -167.8 mV and for Na^+ -103 mV, using the molarities quoted above.

REFERENCES

Bowling, D. J. F. and R. M. Spanswick 1964. *J. Exp. Bot.* **15**, 422–17.
Spanswick, R. M. and E. J. Williams 1964. *J. Exp. Bot.* **15**, 193–200.

4.3 Potassium uptake by excised roots
Based on information given by Professor D. A. Baker, Wye College
cf. Expts 4.4, 5.3, 7.4, 9.2 & 9.5

To avoid a gradual increase in concentration of the KCl solution by evaporative loss, dishes must be covered except for the brief moments when solution is withdrawn.

It is necessary to aerate the solutions gently in order to avoid localised concentration gradients and to provide the root sections with an aerobic environment.

About 5 g of well washed, 1-cm-long, apical pieces of primary roots from maize seedlings grown for 2 weeks in trays lined with moist filter paper are used. The excised roots are placed in 50 cm³ of a 10^{-1} mol m^{-3} (100 μM) KCl solution. Every 30 min 5 cm³ of the KCl solution are withdrawn for a $[K^+]$ determination in the flame photometer. The results are plotted against time.

A parallel experiment is carried out with a solution of 10^{-1} mol m^{-3} (100 μM) KCl plus 10^{-3} mol m^{-3} (1 μM) carbonylcyanide-m-chlorophenylhydrazone, an inhibitor of oxidative phosphorylation (cf. Expt 7.4).

4.4 Potassium and water uptake by plants
Based on information given by Professor D. A. Baker, Wye College
cf. Expts 4.3, 5.1–5.3, 9.2 & 9.5

Plants must be suitably supported in their vessels, well illuminated and exposed to some air movement. The room should not be too stuffy and rich in CO_2.

Culture vessels can be constructed from inverted polythene bottles (see Fig. B4.4) of 500 cm³ or larger.

Volume measurements must be accurate and the mixing of the made-up solutions very thorough.

Castor bean (*Ricinus communis*) or tomato (*Lycopersicon esculentum*) plants (see Part A) are used, of about 20 cm in height and with a stem of about 1 cm diameter. The plants are grown in polythene culture vessels provided with a drainage tap (see Fig. B4.4). The experiments must be carried out in light and a very gentle air movement. The nutrient solution (see Part C1) is drained and allowed to drip for exactly 1 min. Next, 250 cm³ of 1.0 mol m^{-3} (1 mM) KNO_3 solution are added to the culture vessel and after 30 min this solution is drained into a 250 cm³ volumetric flask (also allowed to drip for 1 min). The volume of distilled water needed to make up the solution to the mark is recorded and represents the volume of solution taken up. A second 250 cm³ of 1.0 mol m^{-3} KNO_3 solution is supplied to the roots and the measurements are repeated after a further 30 min.

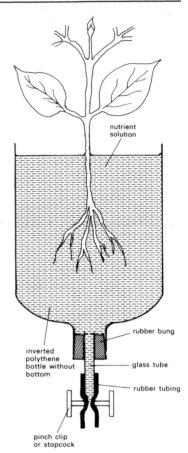

nutrient solution

rubber bung

inverted
polythene
bottle without
bottom

glass tube

rubber tubing

pinch clip
or stopcock

Figure B4.4 Longitudinal section of culture vessel used in Experiment 4.4.

The drained solutions, after being made up to 250 cm³ and very thoroughly mixed, are analysed for K^+ in the flame photometer. Potassium uptake per hour can be expressed in milli-equivalents:

$$2([K^+]_1 \times 0.25 - [K^+]_2 \times 0.25)$$

$[K^+]_1 = 1 \text{ mol m}^{-3}$ and $[K^+]_2 =$ experimental value.

The plants may now be defoliated or detopped and the measurements repeated for another 30 min.

4.5 Histochemical test for potassium
**Based on information given by Dr C. Willmer, University of Stirling
cf. Expts 6.4 & 6.5**

Epidermal strips of *Commelina communis* provide convenient tissue for this test (see Part A). Strips are taken from leaves of plants kept in the dark for 2 h to close stomata and from leaves floating with their lower epidermis in contact with water for 2½ h while illuminated by a 60 W incandescent bulb 40 cm above the dish to open stomata. A fan should blow gently across the dishes to prevent overheating.

Observations as to the state of the stomata should be made prior to the test, as the treatment will close the open stomata.

It is essential that treatments are carried out in dishes surrounded by crushed ice to maintain ice-cold temperatures. During all treatments the tissue should be agitated continuously.

Aeration of the NO_2-evolving mixture will speed up the process.

As NO_2 is evolved, this should be prepared in a fume cupboard and the cobaltinitrite treatment of the tissue only begun after all NO_2 has been evolved.

Stomata that were originally open will have heavy deposits in their guard cells; those that were originally closed very little deposit.

The strips must be rinsed for 2 min in ice-cold distilled water to remove external K^+ before transferring to ice-cold, freshly made sodium cobaltinitrite solution prepared as follows: 20 g cobalt nitrate and 35 g sodium nitrite are dissolved in 75 cm^3 dilute acetic acid (10 cm^3 glacial made up to 75 cm^3 with water). After keeping the epidermal tissue for 10–15 min in the ice-cold solution, it is washed in three changes of ice-cold distilled water for about 2 min until the washings are no longer yellow.

After washing the tissue, it is immersed in a freshly made solution of yellow ammonium sulphide (5 cm^3 Analar $(NH_4)_2S$ plus 95 cm^3 H_2O) at room temperature for 2 min in a fume cupboard and thereafter washed in water to remove external deposits and excess ammonium sulphide. When it is mounted in water under a coverslip, black precipitates will show the K^+ distribution in subsidiary and guard cells.

4.6   Nitrogen determination

Based on information given by Dr J. M. Wilson, University College, Bangor cf. Expts 2.8, 4.8, 4.9, 9.2 & 11.4

The Kjeldahl method can be used for plant material as well as for soils, all of which have to be digested by the method described for phosphorus determinations (cf. Expt 4.9; Table B4.6). The distillation of the digest is carried out in the special Büchi apparatus which distils ammonia titratable with HCl. The use of the Büchi apparatus must therefore be understood and, in addition, attention is drawn to the following:

(1)   After switching on the heating, the 3-way steam tap (A) shown in Figure B4.6 must be in the aeration position, the sample funnel tap (B) closed and the water level allowed to rise slowly until a steady state overflow is established.

(2)   Ten cubic centimetres of a solution of 2 g boric acid in 100 cm^3 water and 3 drops of the green–red indicator (see below) are put into a 50 cm^3 beaker placed beneath the condenser outlet.

(3)   A 10 cm^3 sample to be analysed is put into the sample funnel and run into the still via tap (B) (Fig. B4.6), after which the tap must be closed.

(4)   Next, 10 cm^3 of a solution of 40 g NaOH in 100 cm^3 water, together with a few drops of indicator, are added to the distillation section

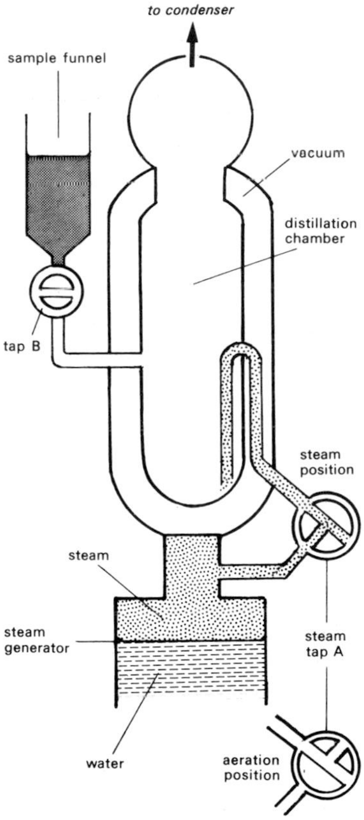

Figure B4.6 Longitudinal section of Büchi still with tap positions indicated as described in the text.

Table B4.6 Approximate protein and nitrogen content of different plant material expressed on a dry weight basis and the amount of fresh weight material needed for analysis.

	% protein on dry wt basis	% nitrogen equiv. on dry wt basis*	g fresh wt needed for digestion
spinach	40	6.0	6.0
pea seed	20	3.0	1.2
maize and wheat	10	1.5	2.4
oak	3	0.45	7.8

* Assuming that there is 15% N in protein.

Calculations

10 mol m^{-3} HCl ≡ 0.14 mg nitrogen

% nitrogen in total fresh wt sample =
(sample titre − blank titre)
　　　　× 5 × 0.14 × 100/ fresh wt

(The figure 5 in the above expression is the dilution factor.)

via the sample funnel. The mixture in the still should be blue, indicating neutralisation of acid by NaOH. The sample funnel is finally rinsed with 5 cm³ distilled water, which are then run into the still.

(5) After setting the steam tap (A) to 'steam', distillation is timed for 2 min *after* vapour begins to pass into the condenser arm. All samples must be treated with the same timing.

It is necessary to flush and rinse the still in the normal way before a new sample is analysed. The distillate collected in the boric acid beaker should be blue and is titrated with 10 mol cm^{-3} (10 mM) HCl to a neutral grey end-point.

INDICATOR

To prepare the indicator, 0.1 g bromocresol green and 0.1 g methyl red in 100 cm³ 95% ethanol are mixed in the ratio of 5 to 1, respectively.

4.7　Nitrogen fixation
**Based on information given by Dr R. Sexton, University of Stirling
cf. Expt 2.8**

The lichen should not be kept overnight after collection, as it deteriorates. The lichen can be replaced by 1 g of root nodules.

After cleaning a portion of *Peltigera canina* thallus collected shortly before use and kept damp, 1.5 g samples of tips of thallus are cleaned of all contaminating material and placed in 25-cm³ conical flasks together with a trace of water to keep the atmosphere damp. The flasks are closed with Suba-Seal (from William Freeman, see Part C2) and some of them covered with aluminium foil for dark treatments. The remainder are exposed to bench lights at about 25 cm distance. One cubic centimetre of acetylene is injected into the flasks by pumping the syringe several times to ensure good mixing; a few flasks, however, are prepared without acetylene and some without lichen.

When withdrawing the syringe after acetylene injection, it must be ensured that the plunger is left down (cf. Expt 2.8).

After 0, 15, 30 and 60 min 1.0 cm³ gas is withdrawn from the flasks containing acetylene and after 90 min from all the flasks including those without lichen and without acetylene. The 1.0 cm³ samples are analysed in a gas chromatograph for the estimation of ethylene production in light and in darkness, expressed as moles per gramme fresh weight per hour.

If the 1.0 cm³ samples cannot be analysed immediately, the syringes containing them may be kept submerged in water in a test-tube until the gas chromatograph is available or the needles may simply be pushed into a rubber bung.

A calibration of the chromatograph is prepared using a standard gas of 1 cm³ pure ethylene and 1 cm³ pure acetylene injected into a 1000 cm³ volumetric flask with magnetically stirred 'flea'. A suitable chromatographic system is a column of Porapak (from Phase Separation, see Part C2) maintained at 65 °C. Samples of 1.0 cm³ from the standard gas mixture in the volumetric flasks are injected into the gas chromatograph. The second marked peak is produced by ethylene and the third, well separated on a chart moving at 4 cm min^{-1}, by acetylene.

If the lichen material is scarce, the experiment can be carried out with 0.5 g thallus samples placed in 10-cm³ conical flasks. The volume of acetylene injected is then 0.4 cm³ and the samples withdrawn for analysis also 0.4 cm³.

REFERENCE

Hitch, C. J. B and W. D. P. Stewart 1973. *New Phytol.* **72**, 509–24.

4.8   Protein estimation by dye-binding
Based on information given by Dr J. F. Farrar, University College, Bangor
cf. Expts 2.5, 4.6, 9.2 & 11.4

Although pigmented material can be used with this method, non-green tissue might prove more suitable.

The NaCl solution is used in order to ensure that proteins adsorbed on to cell debris will be released (cf. Expt 2.5).

Plant material, 0.2 g fresh wt, is ground up in a mortar with 4 cm³ of 0.5×10^3 mol m⁻³ (0.5 M) NaCl at pH 7.0 (10^2 mol m⁻³ (0.1 M)) phosphate buffer (see Part C1). The slurry is centrifuged for 5 min at 1600 g and the supernatant used for analyses. If some debris floats on the surface, it must be filtered off or the supernatant should be removed with a Pasteur pipette.

A calibration curve is prepared using 10–100 μg bovine serum albumin (or any other known protein) in 0.1 cm³ samples mixed with 5 cm³ dye solution (see below). Absorbance is read at 595 nm (5950 Å) and the spectrophotometer is zeroed with 0.1 cm³ blank extraction medium (NaCl at pH 7.0) and 5 cm³ of the dye. After calibration 0.1 cm³ sample extract is mixed with 5 cm³ of the dye for absorbance measurements and the calculation of results.

Average protein levels found in parts of maize plants are:

$$25 \text{ μg g}^{-1} \text{ fresh wt in roots}$$
$$135 \text{ μg g}^{-1} \text{ fresh wt in stems}$$
$$400 \text{ μg g}^{-1} \text{ fresh wt in leaves}$$

DYE SOLUTION

The Coomassie blue method has been chosen for class experiments in preference to the established, but complex, Folin Ciocalteu method.

To prepare the dye solution, 100 mg Coomassie brilliant blue dye are put into 50 cm³ 95% ethanol plus 100 cm³ phosphoric acid made up from 85 cm³ acid plus 15 cm³ water; the solution is then made up to 1000 cm³. The solution must be shaken thoroughly and filtered just before use. It lasts for about 2 months.

REFERENCE

Bradford, M. M. 1976. *Analyt. Biochem.* **72**, 248–54.

4.9   Phosphorus determination
Based on information given by Dr J. M. Wilson, University College, Bangor
cf. Expts 4.6, 4.8, 9.2 & 11.4

Heating must be carried out in a fume cupboard.

For both phosphorus and nitrogen determinations (cf. Expt 4.6), plant material must be digested. About 1 g fresh wt of material, with the weight recorded to four decimal places, is digested in 5 cm³ concentrated N-free sulphuric acid in a digestion flask. Heat, with occasional mixing, until the digest becomes clear (1–3 h). After cooling the contents, they are made up with distilled water to 50 cm³ in volumetric flasks and kept with closed tops in a refrigerator to avoid atmospheric nitrogen absorption.

The spectrophotometer can be calibrated by using 5-, 10- and 15-cm³ samples from a known working standard (see below) in 50-cm³ volumetric flasks filled to about two-thirds with distilled water plus 0.2 cm³ of the digestion acid. One additional flask (without any working standard solution) is also needed. To each of these calibration mixtures are added 2 cm³ of ammonium molybdate reagent and 2 cm³ stannous chloride reagent (see below) and the mixture is made up to 50 cm³ with distilled water. After each addition the contents are mixed thoroughly and allowed to stand for 30 min before the optical density is determined at 700 nm (7000 Å) with water as a reference for zeroing the instrument. A calibration line can now be drawn.

For the phosphorus determinations of the digest samples, 2-cm³ aliquots of digest are pipetted into 30 cm³ distilled water contained in clean 50 cm³ volumetric flasks. Thereafter, 2 cm³ of each of the two reagents are added and mixed as for the standards and the flasks filled up to 50 cm³ with distilled water. After allowing the contents to stand for 30 min, the optical density is determined at 700 nm (7000 Å).

% phosphorus

$$= \frac{(\text{mg P from calibration}) \times (50 \text{ cm}^3 \text{ solution volume}) \times 100}{(\text{mg fresh wt sample}) \times (2 \text{ cm}^3 \text{ aliquot used})}$$

REAGENTS

PHOSPHORUS WORKING STANDARD

A phosphorus stock solution is prepared of 0.4393 mg *dry* KH_2PO_4 1000 cm⁻³ (1 cm³ = 0.1 mg P). A working standard is obtained by diluting this solution 50 times (1 cm³ = 0.002 mg P).

AMMONIUM MOLYBDATE REAGENT

This is prepared from two solutions, A and B:

solution A is ammonium molybdate –
        25 g $(NH_4)_6 Mo_7O_{24} \cdot 4H_2O$ in 200 cm³ H_2O;
solution B is prepared by adding, with cooling (0 °C on ice) and
        mixing, 280 cm³ concentrated H_2SO_4 to 400 cm³ H_2O.

Solution A is filtered into solution B through Whatman No. 1 paper and the mixture is made up to 1000 cm³ with distilled water. The ammonium molybdate reagent is cooled and stored in the dark.

STANNOUS CHLORIDE REAGENT

All glassware must be very clean.

The distilled water used for making up standard volumes must always come from the same source.

The reagent ($SnCl_2 \cdot 2H_2O$) is prepared by adding a solution of 5 cm³ concentrated HCl in 250 cm³ H_2O to 0.5 g stannous chloride in an empty 250 cm³ volumetric flask, and making up to the mark. This solution must be prepared immediately before use.

4.10   Mineral deficiencies
**Based on information given by Professor D. A. Baker, Wye College
cf. Expt 9.1e**

Seedlings of Castor bean (*Ricinus communis*), tomato (*Lycopersicon esculentum*), cabbage (*Brassica oleracea*) or other *Brassica* spp. grown in nutrient solutions (see Part C1) to the first leaf stage are transferred to eight different aerated nutrient solutions, seven of which are deficient in one macronutrient. The plants are allowed to grow at normal temperature and day length. The solutions are made up as shown in Table B4.10b.

Table B4.10 (a) A key to deficiency symptoms.

Symptoms	Element deficient

I   Older leaves affected

(a)   Effects mostly generalised over the whole of the plant, lower leaves dry up and die

if plants light green,
   lower leaves yellow, drying to brown,
   petioles becoming short and slender — **nitrogen**

if plants dark green with red or purple colour appearing,
   lower leaves yellow, drying to dark green,
   petioles becoming short and slender — **phosphorus**

(b)   Effects mostly localised, mottling or chlorosis, lower leaves do *not* dry up, but become mottled or chlorotic with leaf margins cupped or tucked

if leaves mottled or chlorotic, sometimes reddened,
   necrotic spots,
   petioles slender — **magnesium**

if leaves mottled or chlorotic,
   small necrotic spots between veins or near leaf tips and margins,
   petioles slender — **potassium**

if leaves mottled or chlorotic,
   large necrotic spots generally distributed and involving veins,
   leaves thickened,
   short petioles — **zinc**

II   Younger leaves affected

(a)   Terminal buds die, distortion and necrosis of leaves

if young leaves hooked,
   dying back at tips and margins — **calcium**

if young leaves light green at base,
   dying back from base,
   leaves twisted — **boron**

(b)   Terminal buds remain alive but chlorotic or wilted without necrotic spots

if young leaves wilted,
   without chlorosis,
   stem tips weak — **copper**

if young leaves *not* wilted and chlorosis occurs,
   small necrotic spots,
   veins remain green — **manganese**

without necrotic spots,
   veins remain green — **iron**

without necrotic spots,
   veins become necrotic — **sulphur**

Adapted from: American Potash Institute 1948. *Diagnostic techniques for soils and crops*. Washington, DC: American Potash Institute. Bidwell, R. G. S. 1979. *Plant physiology*, 2nd edn. New York: Macmillan.

Table B4.10 (b) Composition of nutrient solutions for mineral deficiency experiment.

	Complete (cm³)	−Ca (cm³)	−S (cm³)	Stock solution −Mg (cm³)	−K (cm³)	−N (cm³)	−P (cm³)	−Fe (cm³)
10^3 mol m^{-3} Ca(NO$_3$)$_2$	10	—	10	10	10	—	10	10
10^3 mol m^{-3} KNO$_3$	10	10	10	10	—	—	10	10
10^3 mol m^{-3} MgSO$_4\cdot$7H$_2$O	5	5	—	—	5	5	5	5
10^3 mol m^{-3} KH$_2$PO$_4$	2.5	2.5	2.5	2.5	—	2.5	—	2.5
1 g 100 cm^{-3} Fe EDTA	2.5	2.5	2.5	2.5	2.5	2.5	2.5	—
10^3 mol m^{-3} NaNO$_3$	—	10	—	—	10	—	—	—
10^3 mol m^{-3} MgCl$_2$	—	—	5	—	—	—	—	—
10^3 mol m^{-3} Na$_2$SO$_4$	—	—	—	5	—	—	—	—
10^3 mol m^{-3} CaCl$_2$	—	—	—	—	—	10	—	—
10^3 mol m^{-3} KCl	—	—	—	—	—	10	2.5	—
Micronutrient stock solution*	2.5	2.5	2.5	2.5	2.5	2.5	2.5	2.5

These constituents are made up to 1000 cm³ with water.
Care must be taken to add the phosphate last and in diluted form to avoid precipitation.

* Micronutrient stock solution contains: 2.860 g 1000 cm^{-3} H$_2$BO$_3$
0.220 g 1000 cm^{-3} ZnSO$_4\cdot$7H$_2$O
0.079 g 1000 cm^{-3} CuSO$_4\cdot$5H$_2$O
1.015 g 1000 cm^{-3} MnSO$_4$
0.090 g 1000 cm^{-3} H$_2$MoO$_4\cdot$H$_2$O

Deficiency symptoms developing in the plants should be compared with those listed in Table B4.10b.

At weekly intervals, shoot and root lengths are measured, leaf number and size determined, internode number and length recorded and any deficiency symptoms noted. Data can be presented as graphs of growth patterns with time.

4.11 Ion uptake and accumulation in light and darkness
Based on information given by Professor J. Barber, Imperial College of Science and Technology

Experiments involving the use of radioactive material must be carried out according to the safety regulations in force. All radioactive waste, including plant material, must be disposed of immediately after the conclusion of experiments.

Ten cubic centimetres of *Chlorella* cell suspension (see Part A) are delivered from a 10 cm³ pipette into sample tubes (8 × 2.5 cm), one of which is blacked out with polythene. To each suspension, 100 mm³ of ^{36}Cl stock solution (see below) are added from a Hamilton syringe and the sample tubes placed securely in a water-bath with shaker at 25 °C. The time of addition of the isotope is noted. Illumination from one side by a 150 W quartz iodide slide projector bulb is provided, the other side of the water-bath being covered with aluminium foil serving as reflector.

All operations must be carried out above suitable trays in case of any spills, and everything that has been in contact with the radioactive substance must be collected in special containers for specialised treatment or disposal.

Samples of 1 cm³ are removed with disposable plastic syringes after about 5 min and thereafter at 10-min intervals for 90 min. Good time-keeping is necessary. The samples are thoroughly mixed with about 10 cm³ distilled water, then filtered under suction through Whatman glass-fibre discs (GF/C 2.5 cm diameter) and *thoroughly* washed with more distilled water while still under suction. A Millipore filtration set (cat. no. XX 1002500) (see Part C2) is ideal for this work and fits a 125 cm³ Büchner flask and funnel.

The filter discs are placed on aluminium planchettes (from Sigma Chemicals, see Part C2; 2.5 cm) and carefully dried under an infra-red lamp, after which they can be glued on to the planchettes, labelled and kept for counting. Six empty filter discs on planchettes are counted for background radiation and three total counts are made with 100 mm^3 of ^{36}Cl-containing algal suspension delivered directly on to filter discs glued to a planchette. A gas flow counter in the Geiger mode is most suitable. If other counters are used, the initial dosage may have to be increased.

SAMPLE CALCULATION

Results should be expressed as quantities of chloride transported in unit time across unit area of cell surface.

SURFACE AREA

Chlorella cells can be assumed to be spherical with a radius of 5×10^{-4} cm. Surface area $= 4\pi r^2$; since a 1 cm^3 sample is assumed to contain 10^7 cells, the total surface area will be 31.42 cm^2.

COUNTS MIN^{-1}

Counts min^{-1} must be converted to moles of chloride: 100 mm^3 of a 0.35 mol m^{-3} chloride solution contain $0.35 \times 10^{-3} \times 10^{-6} \times 100$ moles of Cl$^-$ and therefore

$$1 \text{ count min}^{-1} = \frac{0.35 \times 10^{-7}}{\text{mean counts min}^{-1} \text{ per 100 mm}^3} \text{ moles Cl}$$

For example:

$$\text{(total count} - \text{background)} = 4644.5 - 13.1 \text{ counts min}^{-1}$$
$$= 4631.4 \text{ counts min}^{-1}$$
$$1 \text{ count min}^{-1} = 7.56 \times 10^{-12} \text{ moles Cl}^-$$

RATE OF UPTAKE

Uptake rate can be determined from the slope of the uptake rate curve over the first 5 min. For example, assuming slope is 21.04 count min^{-1} per minute, and converting minutes to seconds, the rate of uptake would be:

$$\frac{7.56 \times 10^{-12} \times 21.04}{60} \text{ moles Cl}^- \text{ s}^{-1} \text{ aliquot}^{-1}$$
$$= 2.65 \times 10^{-12} \text{ moles Cl}^- \text{ s}^{-1} \text{ aliquot}^{-1}$$

For the 1 cm^3 aliquot the total cell surface was 31.42 cm^2 and therefore the initial rate of uptake in light would be:

$$\frac{2.65 \times 10^{-12}}{31.42} \text{ moles Cl}^- \text{ s}^{-1} \text{ cm}^{-2}$$
$$= 8.43 \times 10^{-14} \text{ moles Cl}^- \text{ s}^{-1} \text{ cm}^{-2}$$

$$= \left(\frac{1}{2.13}\right) \ (4.0 \times 10^{-15}) \ \left(\frac{1 - \exp - 2.13}{0.35 \times 10^{-6}}\right) \ cm \ s^{-1}$$

$$= \left(\frac{1}{2.13}\right) \ (4.0 \times 10^{-15}) \ \left(\frac{1 - 0.12}{0.35 \times 10^{-6}}\right) \ cm \ s^{-1}$$

$$= \frac{4 \times 10^{-15} \times 0.88}{2.13 \times 0.35 \times 10^{-6}} \ cm \ s^{-1}$$

$$= 4.7 \times 10^{-9} \ cm \ s^{-1}$$

where $Z = -1$ (number of electrons involved), $F = 96\,000$ Coulomb equivalent^{-1} (Faraday constant), $E = -50$ mV (electropotential difference across cell membranes), $R = 8.3$ J K^{-1} (gas constant), $T = $ temperature in degrees Kelvin, $K = $ degrees Kelvin, and $[Cl^-]_0 = 0.35 \times 10^{-6}$ moles cm^{-3}.

5 Water relations

5.1 Sap exudation, guttation and root pressure
By Professor H. Meidner, University of Stirling
cf. Expts 5.2 & 5.3

Exudation from detopped plants has been dealt with in Experiment 4.1.

GUTTATION

For guttation droplets to collect, the atmosphere surrounding the leaves must be of a high vapour density. Plastic hoods, glass bell-jars or beakers large enough not to touch the shoot and which can be removed without disturbing the leaf margins must be placed over the plants. A good contact between the covers and the bench is desirable but it need not be an airtight seal. As the covers are likely to become fogged by condensation on the inside, it is desirable that they can be removed for closer observation of the guttation droplets without disturbing these.

As an instructive parallel experiment it is recommended to 'water' some of the dishes containing seedlings with 5×10^2 mol m^{-3} (0.5 M) sucrose or mannitol solution – guttation will not occur.

Another form of exudation, namely guttation from intact rooted plants, can complement these studies. For exudation and guttation a differentially permeable membrane, effectively the whole root, must be present, as these processes are due to an osmotic flow from high to low water potentials; hence a cut shoot cannot guttate as there is no differentially permeable membrane separating it from the solution. All cereal seedlings are foolproof material, young tomato plants and many other seedlings will guttate, both tomato and maize plants up to about 30 cm in height work very well. The composition of exudate from detopped plants and of guttation droplets is likely to differ, as the guttation droplet is excess xylem sap out of which most of the solutes have already been removed by the surrounding tissues. Guttation experiments should be combined with microscopic studies of hydathodes (see below).

ROOT PRESSURE

The measurement of root pressure can be performed with U-tube manometers or with sensitive pressure transducers attached at A of a U-tube manometer (see Fig. B5.1). The use of mercury in the manometer presents unnecessary difficulties; an alternative is bromoform (4 times as heavy as water) coloured with a dye soluble in it, e.g. Janus red, but several others can be found by trial and error. Plants must have somewhat woody stumps of 0.5–1.0 cm diameter. *Pelargonium zonale*, *Ricinus communis* and *Xanthium strumarium* are suitable (Part A).

RELEVANT ANATOMICAL STUDIES

For the preparation of material to observe hydathodes and spiral thickening of vessel walls in leaves by clearing in lactophenol see Part A, Hydathodes.

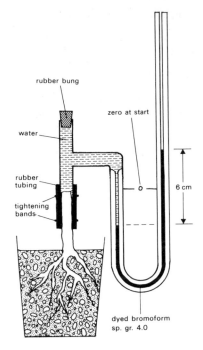

Pressure is proportional to the *difference in levels of the manometer liquid in the two limbs*. If the liquid is bromoform with a sp. gr. of 4.00, a difference in level of 6 cm would be equivalent to a pressure of 24 cm water column.

Figure B5.1 Longitudinal section of manometer assembly for the measurement of root pressure.

5.2 Water inflow, translocation and water vapour loss
By Professor H. Meidner, University of Stirling
cf. Expts 4.4, 5.1, 6.1–6.3 & 9.1b–d

Reference is made here to standard, classical experimental arrangements. Because these are well known and descriptions can be found elsewhere (Meidner & Sheriff 1976; most standard texts), only practical hints and suggestions for variations in the procedures and aims of the experiments will be mentioned.

On the basis of today's interpretation of water movement into plants, i.e. from high to low water potentials, the term 'inflow' has been chosen, rather than 'uptake' or 'absorption', in order to emphasise that the energy causing water to move into the plant resides in the outside solution with a higher water potential than that of the plant. This usage is valid both for osmotic inflow into roots, predominating in seedlings, and for 'transpiration stream'-inflow when positive hydrostatic pressures are first reduced to zero and thereafter become negative so that plant water potentials fall far below soil water potentials.

USE OF CUT SHOOTS

Species with round woody stems are most easily handled.

The choice of leafy woody branches will vary with locality and therefore specific recommendations cannot be made. Branches must be prepared as described in Part A, if good results are to be obtained. The most convenient arrangement is to use a 100 cm³ filter flask on to the side arm of which a 15–20-cm-long glass capillary is attached by means of a 3-cm-long piece of rubber or plastic tubing. The capillary tube should have a downward bend at the free end (flame-polished), which dips into a beaker of water. The flask must be filled to the brim with water. When

Vaseline must not be used to obtain airtightness; the twig should be a good fit in the bore of the bung in the first place and compression of the bung when pushed into the neck of the flask will do the rest. If a matrix has to be used to achieve an airtight fit, a quick-setting silicone rubber may be tried.

Fitting the twig in the rubber bung and then in the flask while the whole assembly is held under water in a sink avoids trapping air bubbles in the system.

Instead of introducing a new air bubble into the capillary tube for repeated measurements, it is very convenient to use a small syringe fitted as shown in Figure B5.2a for resetting the bubble or a free meniscus of the water column in the capillary.

pushing the rubber bung, with the twig already fitted, into the neck of the flask, the rubber will make an airtight joint around the branch and water will shoot along and out of the capillary tube expelling all air. The bent end of the capillary must remain under water, otherwise air will be sucked back into the flask when the hold on the rubber bung is released. When the apparatus has been assembled, only the smallest of air bubbles, if any, should remain at the base of the bung and the prepared end of the twig must dip 2–3 cm into the water inside the flask. To introduce a 0.3- to 0.6-cm-long air bubble into the capillary, the bent end is lifted free of the water in the beaker, dried with paper tissue and replaced in the water when air has entered the flame-polished end of the tube. Measurements of water inflow rates are now readily made, and are best at steady states.

EXPERIMENTAL VARIATIONS

By varying light supply, atmospheric humidity, air movements or carbon dioxide content of the air, the effects of these factors can be studied. It is instructive to combine some of these with measurements of stomatal conductance by $CoCl_2$-paper (see Part C1) or simple porometer (see Part C1 and Expt 6.1) and to compare the effects of these factors on stomata with their effect on the rate of water inflow into and vapour loss from the shoot and an atmometer (see Part A). After measurements have been made in free air, it is also instructive to place transparent and black

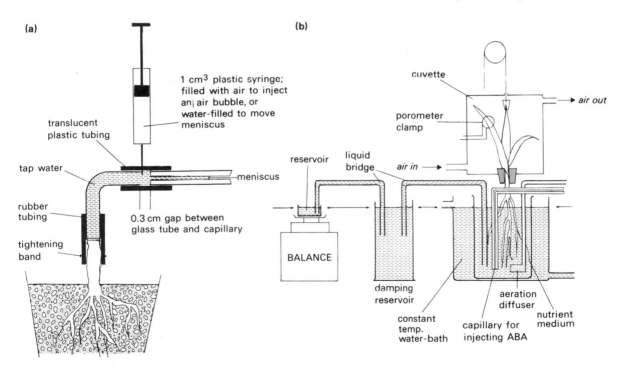

Figure B5.2 (a) Longitudinal section of apparatus suitable for introducing and repeatedly setting an air bubble, or the meniscus in a capillary, for the measurement of rates of sap exudation from the stump of a detopped plant. This diagram shows the root in soil as for Experiment 5.3. For Experiment 5.2 the roots are in nutrient solution. (b) Longitudinal section of apparatus for measuring water inflow rates into rooted plants by weighing. The apparatus can be used also for measuring leaf extension growth, leaf conductance, water vapour loss and the absorption lag. The cuvette surrounding the plant is required only for experiments involving the measurement of water vapour loss.

For the ABA treatment, it is best to inject sufficient of a stock solution to obtain 5×10^{-2} mol m^{-3} (50 μM) in the flask, mix by swirling and continue measurements. Water inflow rates will gradually decline when the ABA reaches the leaves.

The absorption lag will be most marked when a turgid plant is illuminated and exposed to a gentle air current.

By cutting off the root system after measuring the absorption lag, it can be demonstrated that root resistance to water inflow is considerable; water inflow rates will speed up after removal of the roots. That is the reason for using rooted plants for this experiment – with cut twigs the absorption lag is much reduced or practically absent.

The nutrient solution should be aerated for better growth. Additions of ABA to the nutrient medium will allow a study of its effects in an intact rooted plant. Changing the temperature around the nutrient medium container will make it possible to measure the effect of temperature on water inflow rates.

plastic bags over the leaves, especially if this is combined with stomatal tests. By replacing the water with 5×10^{-2} mol m^{-3} (50 μM) abscisic acid (ABA) in the flask and noting the time required to effect a change in the rate of water inflow, translocation rates can be estimated and the effect of ABA on stomata can also be assessed (see p. 93).

THE ABSORPTION LAG

For the measurement of the absorption lag, small, rooted plants must be used. *Coleus blumei*, *Pelargonium zonale* or any other species the cuttings of which root readily and grow well in nutrient solution are suitable. The stem must be held airtight in a split rubber bung suitably fashioned for it.

The assembly described for experiments with cut twigs can be put on a top-loading balance to measure transpiration by weighing, and water inflow rates by timing the receding meniscus over a known distance. The syringe refill system must be used.

VARIATION OF THE APPARATUS
(Based on information given by Dr W. J. Davies, University of Lancaster.)

The assembly shown in Figure B5.2b is more versatile than that described above. In this version water inflow rate is measured by weighing, and transpiration rate by psychrometer, using wet and dry thermocouples or other psychrometers in the stream entering and leaving the cuvette surrounding the plant. For measurements of gains in weight of CaCl$_2$ due to water vapour absorption, dried air must be used to enter the cuvette and calcium chloride tubes fitted into the outgoing airstream. Simultaneous porometer (see Part C1) and leaf growth measurements (see Expt 11.9) can be made if arrangements are made as shown in Figure B5.2b.

REFERENCE

Meidner, H. and D. W. Sheriff 1976. *Water and plants*. London: Blackie.

5.3   Osmotic effects of external media on solution uptake by root systems
Based on information given by Professor D. A. Baker, Wye College
cf. Expts 3.2, 4.3, 4.4 & 5.1

Removal of the tissue around the xylem must be done neatly without cutting into the xylem. The rubber tubing must be a watertight fit, and *must* be pushed over the stump and the glass capillary when both are *wet*. If the rubber tube is filled with water after pushing it over the stump, air bubbles should not lodge between the stump and the glass capillary (see Figure B5.2a for details). The syringe remains in place throughout the experiment.

Detopped plants, with a 3–5-cm-long stump, growing in 1.0 mol m^{-3} (1 mM) KNO$_3$ solution are used (such plants may be available from Expt 4.1). The tissue surrounding the xylem cylinder must be neatly trimmed off, exposing the clean xylem. A piece of tightly fitting rubber tubing about 3–4 cm long is pushed over the xylem and filled with water. The glass capillary tube shown in Figure B5.2, with its flame-polished end near its bend, is now inserted into the water-filled rubber tube so that the capillary will fill with water. A small air bubble can be introduced from a 1 cm^3 plastic syringe and its rate of progress measured with a stop-watch.

When a steady-state rate has been established, the 1.0 mol m^{-3} (1 mM)

In a mannitol medium roots will function osmotically. In a KNO_3 medium the effects of ion transport will alter the osmotic pattern (cf. Expt 3.2).

KNO₃ solution is replaced with a 10 mol m⁻³ (10 mM) mannitol solution and the rate of exudation measured again till a new steady state is established. After replacing the mannitol with the original 1.0 mol m⁻³ (1 mM) KNO₃ solution, the experiment can be repeated with a 10 mol m⁻³ (10 mM) mannitol solution.

RELEVANT ANATOMICAL STUDIES

Thin hand-cut sections of stems of detopped plants mounted in saffranin (see Part C1) will allow the speedy recognition of the pattern of xylem and sclerenchyma distribution in the stem. If the section is thin enough, the unstained phloem tissue can also be identified. Once the anatomy of the stem has been observed microscopically, it will be possible to determine by using a hand lens from which tissue of the stump exudation occurs (cf. Expt 5.8).

5.4 Tissue water potentials, ψ_{tissue}
By Professor H. Meidner, University of Stirling
cf. Expts 3.2, 5.7 & 5.8

(A) BY PSYCHROMETER

If fully turgid tissue is used to begin with, results appear uninteresting when plotted because such tissues have a water potential of zero. It is therefore recommended to use tissue that has lost some turgor.

The water potentials of tissues (ψ_{tissue}) can be determined in calibrated vapour pressure *psychrometers*. Care must be taken that there is no surface moisture on the tissue, especially where it has been cut.

Tissues should be handled with tweezers and uniformly blotted in a gentle manner.

(B) BY IMMERSION IN BALANCING SOLUTIONS

All solutions must be labelled, not on the lid of the dish but *on its base*; irregular results are usually due to mixing up the dishes.

Discs 0.1 cm thick or less from No. 8 cork-borer cores (1.2 cm internal diameter) of storage roots or tubers, leaf discs, hypocotyl sections of cress (*Lepidium sativum*) or cereals are suitable materials if changes in weight are to be measured. All materials must be cut with a new, sharp blade. They must *not* be washed, but instead the tissue should be gently blotted with filter paper before each weighing (otherwise herein lies a serious source of error).

Measurements should be carried out in rotation, and reasonable time-keeping is required for each treatment; this will occupy students continuously.

If the dimensions of strips of tissue are to be measured, good rulers, or wooden or metal callipers with vernier gauges should be used. Strips must be no less than about 5 cm long and about 0.5 cm thick and wide.

The solute potentials of molar sucrose solutions are shown in Figure C3.

Graded solutions of sucrose or mannitol are needed and readings should be taken every 30 min for 2½ h from the start, ideally until equilibrium is reached and measurements do not change between consecutive readings. A zero-ψ_s treatment with tap water must be included.

For the zero ψ_s treatment distilled water is not recommended; tap water should be used (see Part A). In tap water and solutions of low ψ_s, equilibrium will be reached in 2–3 h. At low ψ_s cells may plasmolyse and the heavy solutions will then enter the tissue making it heavier with time; hence the 'kink' (cf. Expt 5.5d).

Results are expressed as percentage changes based on the original measurement (=100%) and plotted against time. This will result in as many curves as there were treatments. A second single plot should be constructed from the percentage changes at equilibrium in each solution (or after 2½ h) against the ψ_{solute} of the solutions. Where this curve cuts the x-axis, the $\psi_{tissue} = \psi_{solute}$ of the bathing solution. If the curve has a 'kink' below the x-axis, the 'kink' will be in the region of ψ_{solute} of the tissue sap, because once incipient plasmolysis has been reached in the bathing solutions, very little further shrinkage of the tissue occurs and weight changes may indeed be the reverse of those in weaker solutions (cf. Expt 5.5d).

Note that the y-axis will have its zero value in its centre; there are increases (+) and decreases (−) to be plotted. The x-axis has its origin at zero at the centre of the y-axis.

The solute potentials of molar sucrose solutions are shown in Figure C3.

It will be convenient to keep the labelled sample bottles in holes drilled into wooden blocks, or pressed into plasticine.

A good check on both the technique and the accuracy of preparing the range of solutions is to plot the initial refractive indices of the solutions against their molarity. A *precise* linear relationship should be obtained.

Another method for determining tissue solute potentials which is based on the same principle, but does not require a refractometer, is Shardakoff's dye method (for details see Meidner & Sheriff 1976).

(C) BY REFRACTOMETRY

(Based on information supplied by Dr E. S. Martin, Plymouth Polytechnic.)

The preparation of tissue and of solutions is essentially the same as for Experiment 5.4b. However, only sucrose solutions may be used. Exactly 1 cm³ of each graded solution is placed in a sample bottle (4 × 1.5 cm) with a plastic lid which must be kept closed at all times when not in use to prevent evaporation. One drop of solution is removed from each bottle in turn and its refractive index measured.

Discs of tissue from cores of storage roots, tubers, etc. are prepared, but they must not be allowed to dry out, nor may they be washed; they should be gently blotted with paper tissue and then incubated for 2 h in the sample bottles containing the solutions.

After 2 h, when the tissue can be assumed to have come into equilibrium with the solution, the contents of each bottle are gently mixed and one drop is removed for the determination of the refractive index. Results are plotted as differences in refractive index between the original and the final solution against the molarities of solutions used. The ψtissue is read off where no change in refractive index has occurred.

REFERENCE

Meidner, H. and D. W. Sheriff 1976. *Water and plants*. London: Blackie.

5.5   Solute potentials of sap (ψs)
By Professor H. Meidner, University of Stirling
cf. Expts 3.2, 5.7 & 5.8

(A)   BY 50% PLASMOLYSIS–DEPLASMOLYSIS

Suitably thin pieces of tissue, e.g. epidermal strips or thin sections of beetroot (*Beta vulgaris*), are placed in a series of graded mannitol solutions as for Experiment 5.4b. Examination of the tissue pieces under 100× magnification after 3 min will indicate in which solutions most of the cells have become plasmolysed and in which plasmolysis has occurred in few cells or none.

Counts of the total number of cells and of plasmolysed cells can then be made preferably in several 400× magnification fields and as rapidly as possible, so that the examination of the different tissues is completed in about 10 min. Plotting the percentage of plasmolysed cells against molarities of the mannitol solutions used will result in a sigmoid curve from which the molarity of solution which would cause 50% plasmolysis can be determined. The solute potentials so determined are those of cell sap at incipient plasmolysis, i.e. at zero turgor.

(B)   BY PSYCHROMETRY

The ψs of the sap released from frozen and thawed tissue can be determined by speedily placing a drop of the sap on filter paper discs inserted in the chamber of the calibrated vapour pressure psychrometer. It should be noted that such sap may be diluted with apoplastic water (see

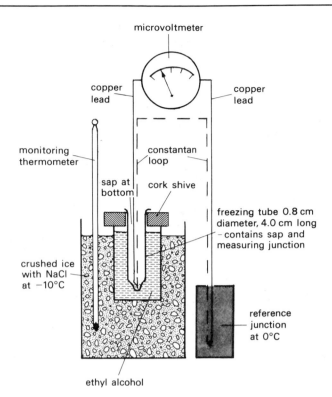

microvoltmeter

copper lead

copper lead

monitoring thermometer

constantan loop

sap at bottom

cork shive

freezing tube 0.8 cm diameter, 4.0 cm long – contains sap and measuring junction

crushed ice with NaCl at −10°C

reference junction at 0°C

ethyl alcohol

Figure B5.5a Longitudinal section of assembly for the cryoscopic determination of ψs of plant sap: the 0 °C reference temperature is maintained by an ice/water mixture kept in a thermos flask with a monitor thermometer.

Care must be taken when supercooling the sap especially since it is colloidal. The process must be slow which will be indicated by a smooth and gradual deflection of the microvoltmeter needle. If the deflection is momentarily interrupted, this shows that the sap has frozen without supercooling and the measurement must be repeated with a *new* drop of sap. Once sap is frozen, its osmotic properties are disturbed.

If sucrose is assumed to be the main component of the sap, the value of − 2.26 in the calculation must be replaced by that relevant to a 1 *molal* sucrose solution and the ΔFP determined for standard solutions as in commercial osmometers. The same considerations apply if solutions of ionising solutes are assumed to be the main constituents of the sap. (cf. p. 60).

Expt 5.7), and the bulk of expressed sap should not be allowed to warm up so that chemical changes are kept to a minimum.

(C)   BY CRYOSCOPY

Sap obtained as for method (B) can also be used in commercial osmometers that have been calibrated with standard solutions. However, the use of a home-made cryoscopic assembly using a thermocouple and calibrated microvoltmeter (see Part A) as shown in Figure B5.5a is strongly recommended because at the same time as measuring the depression of the freezing point it demonstrates the release of the latent heat of fusion of water in a most impressive way.

Once a drop of sap obtained as described above (B) has been placed in the freezing vessel, it must be supercooled by a standard amount. The thermocouple junction in the freezing vessel is then very slightly disturbed. This will cause instantaneous 'setting' of the sap, accompanied by the release of the latent heat of fusion of water, indicated by the instantaneous reversal of the movement of the needle of the microvoltmeter which will come to rest temporarily at the true freezing point temperature of the sap. After a few seconds the frozen sap will cool down again and the microvoltmeter deflection will resume its original direction.

The depression of freezing point (ΔFP) of a solution of a non-ionising solute such as glucose or mannitol (not sucrose) is a measure of its ψs because a solution of 10^3 mol in 1 m^3 of H_2O (1 *molal*) has a ΔFP of 1.86 °C; hence:

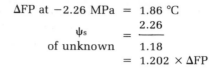

$$\Delta FP \text{ at } -2.26 \text{ MPa} = 1.86 \text{ °C}$$

$$\frac{\psi_s}{\text{of unknown}} = \frac{2.26}{1.18}$$

$$= 1.202 \times \Delta FP$$

(D)   BY A DENSITY GRADIENT METHOD
(Based on information given by Dr E. S. Martin, Plymouth Polytechnic.)

In principle, density gradient methods are weighing methods. This particular 'weighing' method, however, measures *solute* potentials of sap (ψ_s), whereas in Experiment 5.4b the 'weighing' method measures the *water* potential of the tissue (ψ_{tissue}). This distinction is most important.

The preparation of a density gradient column is a skilled operation. Sucrose solutions (see Part C1) graded in steps of 1×10^2 mol m^{-3} (0.1 M) over the range 3×10^2 to 1.5×10^3 mol m^{-3} (0.3–1.5 M) are prepared and successive solutions alternately stained with either methylene blue or neutral red. For different tissues different ranges and steps may be required.

Ten cubic centimetres of the strongest (1.5×10^3 mol m^{-3}; 1.5 M) solution are pipetted into the bottom of a 100 cm^3 measuring cylinder which rests on the bench inclined at an angle of 30° to the horizontal. It is best to hold the cylinder in this position in a retort-stand clamp. Next, the 1.4×10^3 mol m^{-3} (1.4 M) or next weaker solution is very gently run on top of the 1.5×10^3 mol m^{-3} (1.5 M) solution by placing the tip of the pipette close to the surface of the 1.5×10^3 mol m^{-3} (1.5 M) aliquot. This is repeated with all the solutions and finally the cylinder is *slowly* brought back into the upright position. If the layers do not remain distinct, addition of the solution of lower concentration was too fast. Begin again!

Tissue sections 0.1–0.2 cm thick and 0.1–0.2 cm square are prepared from discs as for Experiment 5.4b using sharp razor blades to avoid crushing the cells. A pretreatment to establish equilibrium between ψ_{tissue} and ψ_s of solution is required as indicated next.

Three tissue sections are placed in each of a series of sucrose solutions in steps of 1.0×10^2 or 2.0×10^2 mol m^{-3} (0.1 or 0.2 M) over the range 0.0–10^3 mol m^{-3} (0.0–1.0 M) and kept in these solutions for 1 h so that near equilibrium can be established.

For the determination of the solute potential (ψ_s) curve, one tissue disc is removed from the strongest bathing solution (10^3 mol m^{-3}; 1.0 M) with a camel-hair brush, gently placed near horizontally just below the surface of the density gradient system at the top of the cylinder, and released. After *exactly* 60 s the 'height' at which the tissue floats is read off *from the base* in cubic centimetres in terms of the cylinder graduations. This is repeated with tissue sections from the other incubation media in order of decreasing molarity.

After averaging results of 'heights' obtained with the three replicate discs from each treatment, the means are plotted against solute potentials (ψ_s) of the pretreatment incubation media. The curve will have two slopes with a fairly distinct break point (see Fig. B5.5b). The first slope will be due to increases in density of the tissue since the apoplast has been penetrated by the denser-than-water incubation media. (This effect is offset to some degree by loss of water from the turgid tissue – hence the

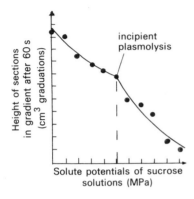

Figure 5.5b Example of curve obtained with the density gradient method for estimating ψ_s of tissue cell sap. The graduations of the measuring cylinder are used to record the 'height' at which tissues float – they are therefore entered as 'cm^3'.

Tissues with airspaces are to be avoided. Potato, beetroot, artichoke, carrot, oat coleoptiles and cress hypocotyls are recommended.

To avoid damaging the tissue, a small camel-hair brush should be used to place sections on the density gradient.

If an air bubble adheres to a section placed in the density gradient system and the bubble cannot be dislodged, another section must be introduced.

gentler slope.) The steeper slope is due to the spaces *between cell walls and plasma membranes* having filled with the denser incubation media which caused plasmolysis. The break point in the curve is therefore a fair measure of the ψ_{solute} of the sap at incipient plasmolysis (cf. Expt 3.2 & 5.4b) when ψ_p is zero.

REFERENCE

Nabors, M. W. 1973. *Am. Biol. Teacher* **35**, 463–9.

5.6 Osmotic adjustment
By Professor H. Meidner, University of Stirling
cf. Expt 9.1e

(A) WHOLE PLANT

A comparison should be made of the solute potentials (ψ_s) determined by 50% plasmolysis (cf. Expts 3.3 & 5.5a) of epidermal cells, some from plants grown in well watered soil and some from plants treated for 3 days with 5×10^2 mol m^{-3} (0.5 M) KCl solution supplied to the soil. Plants of *Saxifraga tomentosa* are recommended as their leaves can readily be peeled and have large pigmented epidermal cells. The plants should be kept in well ventilated rooms and exposed to good illumination during the 3 day treatment.

Results are best expressed as treatment means obtained from 10 or more plants. The ψ_s of the leaf cells of the plants treated with salt solution will be found to be lower than that of the watered plants.

(B) *ULVA LACTUCA* THALLUS
(Based on information given by Dr J. Collins, University of Liverpool.)

Ulva lactuca (sea lettuce) (see Part A), growing near or in estuaries, is periodically exposed to changing water potentials of its aquatic environment. By osmoregulation of ion fluxes, chiefly K$^+$ and Cl$^-$, the solute potential of the tissue sap can be adjusted to changing salt concentrations of the external medium. Without such a mechanism the water content of the cells would change so drastically that irreversible damage would result.

The first consequence of osmotic shock in *Ulva lactuca* is a change in mean cell volume. This is followed by osmotic adjustment or osmoregulation. To detect these responses, plant material should be conditioned by keeping it for 12–15 h in diluted sea water (3 : 1), undiluted sea water and reinforced sea water (+20 g NaCl in 1000 cm^3).

The mean thickness of the thallus should be determined first for the three conditioned tissues. About 0.05-cm-thick sections of thallus should be cut between polystyrene or pith. These sections, standing on edge, can be used to measure thallus thickness using an eyepiece graticule at 400× magnification (see Part A). Mean cell surface dimensions can best be estimated by counts of cells in three 400× magnification microscope fields, or by using calibrated stage and eyepiece grids. From these measurements mean cell volumes can be calculated for the three conditioned thalluses.

Using the material from the undiluted sea water and treating it sepa-

To achieve uniform but gentle blotting of the tissue, it is recommended to roll a 100 g weight 4 times across the filter paper covering the tissue.

Heating should be carried out in a fume cupboard.

If ^{42}K or ^{86}Rb can be used, the ion influx rates can be determined using culture media containing the tracer, and the efflux rates with labelled tissue in ordinary culture medium.

The procedure is complex and reference is best made to West and Pitman (1967) and Black and Weeks (1972).

rately in the reinforced and in the diluted sea water, the osmotic shock effect on cell dimensions can be measured by the procedures outlined above before osmoregulation achieves an adjustment.

To determine the ion content of the three conditioned plant materials, tissue must first be blotted gently between filter papers. About 0.5 g fresh wt tissue, accurately weighed, are digested in 3–5 cm^3 of concentrated HNO$_3$ on a hot plate. When the tissue has totally dissolved, the volume is made up to 10 cm^3 with distilled water and the potassium concentration measured by flame photometry or absorption spectrophotometry.

Chloride concentration can be measured by titrating a distilled water extract obtained at 80 °C from about 0.5 g fresh tissue, blotted and accurately weighed as for the potassium determination.

REFERENCES

Black, D. R. and D. C. Weeks 1972. *New Phytol.* **71**, 119–27.
West, K. R. and M. G. Pitman 1967. *Aust. J. Biol. Sci.* **20**, 901–14.

5.7   Pressure bomb measurements
By Professor H. Meidner, University of Stirling
cf. Expts 5.4 & 5.5

For the successful insertion of material see Experiment 5.8.

For the accurate determination of prevailing water potentials, the material should be enclosed in a plastic bag immediately after cutting it off the plant and then inserted in the bomb. Care must be taken that veins and other tissues are not damaged or punctured.

Pressure bombs are commercially available but can be made in a departmental workshop provided strong material is used and safety precautions are observed. Woody stems are most easily fitted into the airtight gland of the screw-on lid. Petioles, unless round, are difficult to insert in an airtight manner and herbaceous stems may be unduly compressed. It is debatable whether pressure bombs measure xylem water potential or the water potential of the whole twig and leaf tissue.

Twigs cut from well watered plants kept in the dark and illuminated twigs at various degrees of water stress will give instructive results. Climbing stems are especially suitable because of the large diameter vessels.

If leaves are used, the comparison between pressure bomb and psychrometer measurements is valuable. For class use only the commercially available psychrometers can be recommended. Comparisons can also be made between psychrometer measurements of plant water potential and of solute potential of expressed sap after freezing and thawing (Expt 5.5b) and with the results of the pressure–volume curve technique described in Experiment 5.8.

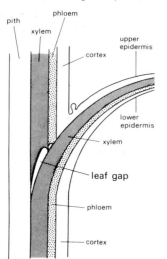

Figure B5.7 Pattern of xylem and phloem strands between stem and petiole forming leaf gap (only half the stem is shown).

RELEVANT ANATOMICAL STUDIES

For the pattern of the vascular system in stems used in the pressure bomb, see Figure B5.7. For studying that of petioles, if leaves are used, transverse, hand-cut sections of the petiole should be prepared. The absence of cambium should be noted and the orientation of the xylem towards the upper and that of the phloem towards the lower epidermis should be explained by reference to Figure B5.7.

5.8 Pressure–volume curves

Based on information given by Dr P. W. Mueller, Imperial College of Science and Technology
cf. Expts 5.4, 5.5 & 5.7

The weighing capsule is a small polythene box into which three filter-paper discs, cut with sharp cork-borers, are fitted; its lid has a hole big enough to allow the petiole to enter. The weight of the capsule is recorded to four decimal places (see Fig. B5.8a).

Fitting the leaf blade into a plastic envelope is good practice preventing evaporative loss. To fit the petiole through the pressure bomb gland from the inside, a hollow probe, wet on the outside, should be used (see Fig. B5.8b). The probe places the petiole through the latex washer of the gland and must be withdrawn once the petiole is gripped by the washer.

The material from which the washer is made is all-important. A very good pliable material can be obtained from Hill's Rubber Co. (see Part C2; type SB 1060, 0.3 cm thick) and some silicone rubber formulations are also suitable. Unless blind washers are used, the bore of the washer should be a tight fit to begin with. With some material it is recommended to use a silicone vacuum grease or rubber grease to lubricate the outside of the probe to avoid splitting or tearing the washer.

Although plant stems and petioles with round cross sections do not present difficulties, there are some which tend to snap when the gland is tightened. In this case a washer with a slightly larger bore should be tried. Petioles with grooves and U-shaped cross sections can be successfully fitted if a little quick-setting glue of the cyanoacrylate type is applied — but, in this case, grease must not be present anywhere. Tissues with large airspaces are not suitable for use in pressure bombs.

This procedure involves repeated pressure applications and absorption of exuded sap into a weighing capsule.

A fully turgid leaf, obtained from a well watered plant and kept with its petiole in water under a plastic hood overnight in the dark, is quickly fitted into a plastic envelope and inserted into the pressure bomb. On the gradual application of a very low pressure (0.01 MPa), such a leaf should exude some sap via its petiole protruding from the pressure bomb. At this stage the weighing capsule (Fig. B5.8a) is fitted over the petiole and the pressure very gradually increased by 0.2 MPa; this pressure shall be called ψ_1. After waiting for 20 min, it can be assumed that no more sap will be exuded at this pressure. The capsule is then removed and placed with its opening facing down until ready to be weighed.

Next, the pressure is slowly reduced by 0.1 MPa and the petiole observed with a hand lens for any sap being reabsorbed. This is followed by a slow increase in pressure until sap just reappears. This pressure shall be called ψ_2, and represents the tissue water potential of the leaf at its present depleted water content; it must be measured accurately. Now the second weighing capsule is placed over the petiole and the pressure raised slowly by 0.2 MPa above ψ_2, allowing for the exuded sap to be absorbed by the filter paper in the weighing capsule which is then removed and weighed together with the first capsule.

This procedure is repeated several times. A plot of $1/\psi$ against the *cumulative* sap yield straightens out towards the x-axis and measurements are continued until three to four points have been obtained on the straight part of the curve; however, the greater the number of points obtained before the curve straightens out, the better for the construction of the graph (see Fig. B5.8c).

Finally, after opening the pressure bomb, the leaf is cut off at the inside of the pressure bomb gland, quickly weighed and then dried in a ventilated oven at 80 °C for 2 h so that the water content at the highest applied

(a)

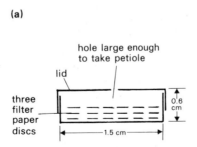

(b)

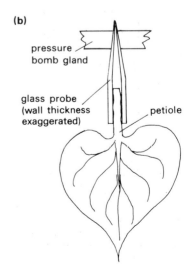

Figure B5.8 (a) Longitudinal section of polythene capsule for collecting sap expressed from the petiole of a leaf enclosed in a pressure bomb. The two halves must be a close fit. (b) Longitudinal section of hollow probe used to insert petioles through the gland of a pressure bomb.

Figure 5.8c Pressure–volume curve used for estimating the solute potential of plant sap of material enclosed in a pressure bomb. The x-axis shows the relative water content (water content at full turgor − volume exuded, expressed as a percentage of the water content at full turgor); the y-axis shows the inverse of pressures applied to exude sap.

A, point at which leaf was fully turgid; B, point of zero turgor, giving value of $1/\psi_s$ at zero turgor; C, last point at which pressure and volume of sap exuded was measured; D, inverse of ψ_s of sap at full turgor; AB, $1/\psi_{leaf}$ at different percentages of water content; BD, $1/\psi_s$ at different percentages of water content; BC, relation of volume exuded to $1/\psi$ (ψ_p = zero). For any given percentage of water content: $1/\psi_{leaf} − 1/\psi_s = 1/\psi_p$.
(d) Modified Höfler diagram relating ψ_p to ψ_{cell} and ψ_s.

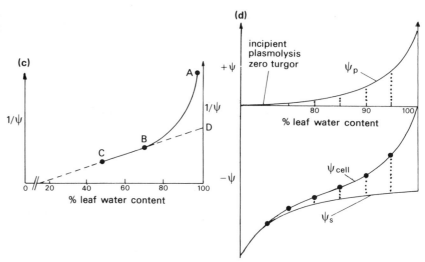

Once the different values have been determined by direct measurement and by extrapolation, it is recommended to construct a modified Höfler diagram as shown in Figure B5.8d to represent the water relations of the tissue.

pressure at the conclusion of the experiment can be determined. By adding the cumulative sap yield obtained during the measurements to the final water content, the initial total leaf water content can be calculated and this should be assumed to represent 100% water content of the fully turgid leaf.

A plot of the results can now be constructed from the reciprocal pressure values and the increments of exuded sap, expressed as percentages of total leaf water content as shown in Figure B5.8c. Line DB, formed by extrapolating the curve from B to point D on the y-axis, represents the ψ_s of the sap, corresponding to the different leaf water contents, whereas curve AB defines the values for the total ψ_{leaf}. The difference between ψ_{leaf} and ψ_s equals ψ_p for any chosen leaf water content.

REFERENCE

Tyree, M. and T. H. Hammell 1972. *J. Exp. Bot.* **23**, 267–82.

6 Stomata

6.1 Stomatal conductance in intact leaves
By Professor H. Meidner, University of Stirling
cf. Expts 4.4, 5.2 & 6.3

Amphistomatous leaves are most suitable for mass flow porometry, e.g. leaves of *Vicia faba*, *Xanthium strumarium*, *Phaseolus vulgaris*, *Pelargonium zonale and Zea mays* (see Part A). *Vicia faba* has leaves from which the epidermis strips easily (cf. Expts 6.3 and 8.5 anatomy) and they are thus also suited for the special microscopic measurements described below. *Xanthium strumarium* leaves are distinctive in that they are heterobaric, i.e. air can pass from stomata in the lower epidermis directly to those in the upper epidermis.

Hypostomatous leaves, practically all those of woody plants, should only be used with transpiration porometers (from Licor, see part C2) or $CoCl_2$-papers (see Part C1) which can, of course, also be used with amphistomatous leaves, but measure *leaf* conductance, not stomatal conductance.

Plant growth conditions (see Part A) must be strictly observed as the wrong time of day in the diurnal rhythm or an accidental incident of water stress makes stomatal responses unpredictable.

Stomatal movements can be monitored in respect of rate and steady states achieved in response to darkness, different quantities and qualities of irradiance (white, red and blue), atmospheric vapour densities, carbon dioxide content of the air and exposure to still air and to different rates of air movement.

Porometer and $CoCl_2$-paper measurements can be combined with microscopic measurements of pore widths in epidermal strips. However, such strips must be taken with the leaf tissue submerged in a few drops of liquid paraffin (sp. gr. 0.86) on a microscope slide (see Part A). This preserves as much as possible the pore widths established in the intact leaf. When strips are taken from tissue in contact with water, hydropassive closure may severely reduce the openings.

It is most convenient to use potted plants. Some are kept in darkness, others in light, some under transparent plastic hoods in the light (to study the effects of $[CO_2]$ and atmospheric vapour density) and some are exposed to gentle air movements from a fan. For rates of closure, plants with open stomata should be transferred to darkness. Transferring plants from one condition to another will set in motion changes in stomatal conductance.

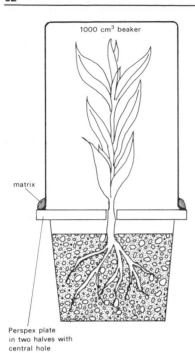

1000 cm³ beaker

matrix

Perspex plate
in two halves with
central hole

Figure B6.3 Longitudinal section of plant chamber formed by inverted beaker on a base plate.

The Perspex plate is best made in two halves with a 1-cm-diameter hole in the centre where the two halves fit together. In this way the plate can be placed on the rim of the plant pot (Fig. B6.3) and a seal around the stem easily effected with a suitable matrix. If large bell-jars are available, they can cover plant pot as well as leafy shoot, but it will be necessary to stretch a plastic sheet over the plant pot rim to prevent soil carbon dioxide from reaching the leaves.

Stripping non-turgid leaves is more difficult than stripping turgid leaves. It is absolutely essential to use a *new* razor blade and to proceed *quickly*.

6.2 Short-period stomatal oscillations
Based on information given by Professor T. A. Mansfield, University of Lancaster
cf. Expt 9.1b & d

Stomatal oscillations, fluctuations or cycles are mainly the result of hydropassive mechanisms and are set in motion most readily in intact plants when fairly sudden changes in either atmospheric moisture deficit, temperature or soil water supply occur while transpiration is vigorous. Six- to eight-week-old *Xanthium strumarium* plants grown in pots are suitable for this experiment. A convenient arrangement for class demonstrations consists of placing the plants under a fairly strong light supply in a fume cupboard which is open by between 5 and 10 cm only, so that it is flushed by a strong air current. Stomatal movements must be monitored with a mass flow porometer such as the Gregory and Pearse resistance porometer (see Part C1) by reading the manometer frequently, or, better, by connecting the porometer via a pressure transducer to a chart recorder.

Once oscillations have been recorded, all leaves except the one to which the porometer is attached should be covered by transparent hoods (e.g. polythene bags). This will reduce or stop the oscillations quickly; removal of the hoods will soon restore them.

6.3 Stomatal opening, carbon dioxide concentration and humidity
Based on information given by Dr W. J. Davies, University of Lancaster
cf. Expt 6.1

Potted plants of *Vicia faba* and *Commelina communis* about 20 cm high are recommended for this experiment. The plants are placed under bell-jars or inverted 2000 cm³ beakers which must be sealed to a Perspex base plate (see below) with Vaseline or plasticine. Illumination from a 100 W bulb or a sunny window should be provided to promote photosynthesis and stomatal opening. After 60 min the covers are removed and it will be seen that the leaves begin to wilt quickly, because stomata become unusually wide open when the leaves are kept under a transparent hood where the concentration of carbon dioxide will be low owing to photosynthesis and the air will be nearly saturated with water vapour. Comparisons with similar plants that are left uncovered will emphasise the point. If possible, it is strongly recommended to take epidermal strips *immediately* the covers are taken off (cf. Expt 6.1). While it is being stripped, the leaf tissue should be kept *in a drop of liquid paraffin* (sp. gr. 0.86), *not in water*, in order to compare pore widths of the treated (covered) and the freely exposed plants. Alternatively, the green leaf tissue itself, mounted in liquid paraffin, allows direct observation of stomatal pores.

6.4 Inhibitors and promoters of stomatal opening
Based on information given by Professor T. A. Mansfield, University of Lancaster
cf. Expt 4.5

For the study of hormone-like regulators of stomatal movements, experiments with epidermal strips are the most useful, but standardisation of the plant material is essential.

Leaves should be used of *Commelina communis* plants that are in the day phase of their endogenous rhythm. It may be advisable to keep the plants in the dark prior to the experiment to reset the rhythm in the correct phase (see Part A). Leaves lying on a wet microscope slide are cut into 0.5-cm-wide strips on either side of the mid-vein. Leaf strips are then 'nicked' with a new razor blade 0.3 cm from their ends in the *upper* epidermis without cutting into the lower epidermis. This can be achieved either while the leaf strip is gently bent over a finger or with the strip lying with its *lower* epidermis downwards on the wet glass slide. After 'nicking', the leaf strip must be replaced on the wet slide with its *lower epidermis facing upwards*. The tissue flap or tab at the end can then be lifted with a pair of forceps (Fig. A7) and pulled gently upwards when the lower epidermis will peel off and can be floated on water. Strips should be divided into pieces about 0.5 × 1.0 cm.

For class experiments only *Commelina communis* is recommended. Other species may well be suitable but they may need different pretreatments and should only be used after a systematic investigation into their responses.

INCUBATION OF *COMMELINA COMMUNIS* EPIDERMAL STRIPS

The medium used for this is a solution of 4.5×10^{-1} mol m^{-3} (0.45 mM) KCl in 10 mol m^{-3} (10 mM) MES buffer (see Buffers, 'Good', Part C1), adjusted with KOH to pH 6.15. After incubation for 3 h in 5-cm-diameter plastic Petri dishes with hypodermal needles inserted through the lids for aeration at about 100 cm^3 min^{-1}, microscopic measurements of the pore widths of 20 stomata can be accomplished in about 5 min (see p. 18).

EXPERIMENTAL TREATMENTS

ABSCISIC ACID (ABA) MIXED ISOMERS (GRADE II: FROM SIGMA CHEMICALS; SEE PART C2)

All treatments are given under uniform illumination and temperature. A range of concentrations from 10^{-1} to 10^{-6} mol m^{-3} (100 μM–1 nM) is recommended to give a graded series of openings. Results should be plotted on a log scale and be correlated with histochemical tests for potassium (Expt 4.5).

The effect of fusicoccin is thought to be due to a stimulation of the proton pump causing increased proton extrusion and potassium influx, as well as changes in cell wall properties.

If fusicoccin is not available, acid-stimulated stomatal opening can be brought about by incubation of epidermal strips in 1 mol m^{-3} (1 mM) HCl for 30 min. However, this low pH effect is at least in part due to the physical destruction of the epidermal cells surrounding the guard cells which alone remain alive. Acid-stimulated stomatal opening cannot depend on the same *metabolic* mechanism as does the effect of fusicoccin.

FUSICOCCIN (FROM FARMOPLANT; SEE PART C2)

Fusicoccin is a spectacular promoter of stomatal opening. It should be used in concentrations of 10^{-4} mol m^{-3} (10^{-7} M) and the strips kept in the dark.

6.5 Histochemical test for chloride

Based on information given by Dr C. Willmer, University of Stirling
cf. Expt 4.5

Raschke and Fellows (1971) recommend floating the tissue for only 5 min in the silver nitrate solution on a microscope slide after which the solution is blotted off the slide. Washing in nitric acid (1.5 cm^3 concentrated acid made up to 100 cm^3) for 10 min is followed by three washes in distilled water. Before examination, the tissue is illuminated with high intensity ultraviolet light under a fluorescence microscope so that the metallic silver particles can be observed.

Leaves with open and closed stomata are collected from *Commelina communis* plants treated as described for Experiment 4.5.

Lower epidermal peels are immersed for 6 h in a solution of 5.0 g silver nitrate in 100 cm^3 water. The tissue is then washed for 30 min in nitric acid (1.5 cm^3 concentrated acid made up to 100 cm^3) followed by three 10-min washes in distilled water. The deposits of silver chloride, which will be more abundant in the guard cells of open than of closed stomata, can be reduced to metallic silver by exposure to ultraviolet light, or several hours of sunlight, until the silver grains appear black when viewed under the microscope.

RELEVANT ANATOMICAL STUDIES
(cf. Dichroic staining, Part A)

A maize variety that has been found to provide leaves that strip fairly satisfactorily is the commercially available F$_1$ hybrid from John Innes (see Part C2). However, it has been reported that many Mexican genotypes are suitable. As with most material used for providing epidermal strips, the growing conditions can make considerable differences. Warm, well watered and adequately illuminated plants appear to have leaves in which the attachment of the epidermis to the mesophyll tissue is less strong than in plants grown under other conditions.

Recommended materials are epidermal peels (see Part A) of *Tradescantia virginiana*, *Commelina communis*, *Vicia faba* and *Zea mays*. Stomata with four and six subsidiary cells will be found in the first two species. *Vicia faba* has stomata without subsidiary cells and guard cells surrounded by epidermal cells with wavy margins, whereas *Zea mays* has typical graminaceous stomata.

Epidermal strips should be stained in neutral red (1:10 000) (see Part C) for 3 min and then mounted in water. The vacuoles of live cells will stain but the cytoplasm and nuclei remain unstained; in damaged cells the vacuoles remain unstained and the cytoplasm and nuclei stain a brownish red. Staining will thus reveal the degree of damage to the epidermal cells. The stain will first be seen in all cells but, if the preparations are left for ½–2 h, it will be noted that the stain becomes concentrated in the guard cells and weaker in the others.

Unstained nuclei are prominent next to the ventral walls of guard cells of ellipse-shaped stomata; they usually lie in the centre next to the ventral wall. If strips are taken from leaves with closed stomata, the guard cell vacuoles will stain a purplish red (pH about 6.4) and, if taken from leaves with open stomata, the stain will be orange red (pH approaching 7.0).

REFERENCE

Raschke, K. and M. P. Fellows 1971. *Planta* **101**, 296–316.

7 Respiration

7.1 The 'respiration train'
By Professor H. Meidner, University of Stirling
cf. Expt 7.2

The inlet and outlet tube levels in each vessel must be arranged as shown in Figure B7.1.

Ensure that all bungs are airtight. Whether air is sucked through from a water tap suction pump or is pushed through by a laboratory air pump, an escape vent with screw clip (see Fig. C2) must be incorporated in order to regulate the flow and provide a slow stream of air bubbling through the flasks containing liquid.

The CO_2 absorber should be strong to ensure that the baryta water in the second flask remains clear.

For the preparation of baryta water, see Part C1; a strong solution may be used in this experiment. For storage, baryta water should be kept in closed bottles with as little air space above the solution as possible to avoid CO_2 absorption and turbidity.

Plentiful material should be used in order to obtain results in a short time.

The measurement of respiratory activity of plant tissue can be carried out at the simplest as well as at the most sophisticated levels. Experiments using Warburg, Gilson or infra-red analysis apparatus are merely mentioned here because of their importance, but they are beyond the scope of this book. The following are examples of qualitative and quantitative experiments at different levels of complexity.

The experimental assembly made from ordinary laboratory glassware is illustrated in Figure B7.1. Into the reaction vessel can be placed any kind of plant material, e.g. germinating seeds, flowers (especially buds), fruit, tubers or yeast suspensions; if leaves are used, the vessel must be darkened or kept at the light compensation point (see Expt. 8.3) — instructive possibilities exist here.

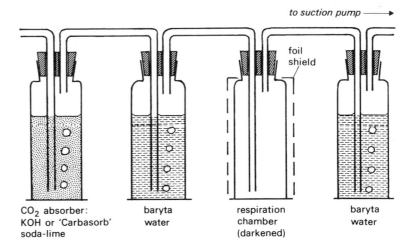

Figure B7.1 Longitudinal section of 'respiration train'.

CO_2 absorber: KOH or 'Carbasorb' soda-lime

baryta water

respiration chamber (darkened)

baryta water

to suction pump ⟶

foil shield

7.2   An improved Pettenkofer tube assembly

Based on information given by Dr R. Phillips, University of Stirling
cf. Expts 7.1, 8.1 & 8.7

The sliding tube inlet to the pressure-regulating vessel should be wetted, or coated with a very thin layer of Vaseline, where it fits into its loosely fitting support bung.

Details of the pressure-regulating vessel (A) and a plant chamber (B) are shown in Figure B7.2c and the entry of the airstream into a Pettenkofer tube is shown in Figure B7.2d.

The airflow through the Pettenkofer tubes should not be much more than about 2000 cm³ h⁻¹. Most air pumps deliver much more than this, but the length of capillary tubing inserted into each Pettenkofer tube offers enough resistance to divert most of the flow to escape through the pressure-regulating vessel (A).

It is good preparatory practice to mark the set of Pettenkofer tubes in their frames at the level to which the baryta water (3 mol m⁻³ (3 mM)) should be filled so that resistance to flow is the same for every tube of the set.

The withdrawal of 1-cm³ samples from the tubes can be done without stopping the airflow. Syringes must be rinsed with distilled water between samplings and after each rinse they should be blown through with compressed air — on no account by mouth.

To obtain measurements in 2 h the baryta water is used at the very low concentration of 3 mol m⁻³ (3 mM). The reaction of baryta water with atmospheric carbon dioxide is rapid. It is essential to reduce exposure of the solution to the laboratory air as much as possible. Therefore, pipetting and titrating must be done rapidly and within the vessels provided. A solution of 150 mol m⁻³ (0.15 M) barium hydroxide is available from Sigma Chemicals (see Part C2).

To obtain reasonable results in the available time for respiration of leaves in the dark and photosynthesis when illuminated, a large leaf area is required. Maize leaves are especially suitable, placed so that they are evenly illuminated and of a length about three-quarters of the height of the vessel.

This classic apparatus employs basically the same methods as the 'respiration train' (Expt 7.1). Indeed, it can be used in the same way, starting with CO_2-free air from an absorption tower if the aim is to measure respiration rate alone. With the improvements here suggested, quantitative measurements are possible and the apparatus may be used as an alternative to infra-red gas exchange equipment. The assembly has been described for four independent measurements. Replicates are essential and the versatility of the apparatus is considerably increased by using six or eight tubes and chambers. The airflow from a laboratory air pump via the pressure-regulating vessel (A) in Figure B7.2a should be such that each of the plant chambers (B) 1, 2, 3 and 4 can receive about 30 cm³ min⁻¹. This rate must be measured accurately later (see below). To obtain this flow, the inlet tube into A can be raised or lowered as required. From the pressure-regulating vessel, the air will pass via a manifold of three Y-pieces into the four plant chambers (B) and hence into the four Pettenkofer tubes (C). Initially, the inlet tube into A must be positioned 5–10 cm below the water level. A quantity of baryta water (see Part C1) can now be put into each of the four Pettenkofer tubes (C) so that the levels are the same in the four tubes in their positions on the wooden frame. This ensures equal resistance to airflow later on. At this stage there should be no inflow of air (indicated by the absence of bubbles in the baryta water), but the air pressure provided by the regulating vessel will keep the baryta water in the limbs of the tubes as shown in Figure B7.2b and this prevents back-siphoning.

From each tube 1-cm³ samples of baryta water are withdrawn with a long-needle syringe and delivered into a capped titration bottle for quick titration with 1 mol m⁻³ (1 mM) HCl using 1 drop of phenolphthalein indicator. The samples must be withdrawn and delivered slowly to avoid air mixing with the solution; the syringe needle and the burette nozzle should enter the hole in the cap of the titration bottle.

Once the baryta water strength has been determined by titration, the airflow from the pressure-regulating vessel is made to enter the plant chambers and the Pettenkofer tubes via the capillary resistances by sliding the escape tube down to about 25 cm below the water level in A. This ensures the correct rate of flow and small bubble formation. The airflow from each tube must be checked with a soap bubble flow meter (see Fig. B7.2e). Every 30 min 1-cm³ samples of baryta water are withdrawn for titration as before.

EXPERIMENTAL MATERIAL

If non-photosynthesising material is used, its weight should be determined for the calculations of respiration rate per gramme fresh weight. With leaves, it is necessary to cut these under water; monocotyledonous leaves are most suitable. Comparable leaf area portions should be used in the different experimental vessels. Darkened vessels are required for respiration rate; vessels illuminated at two light supply levels and a

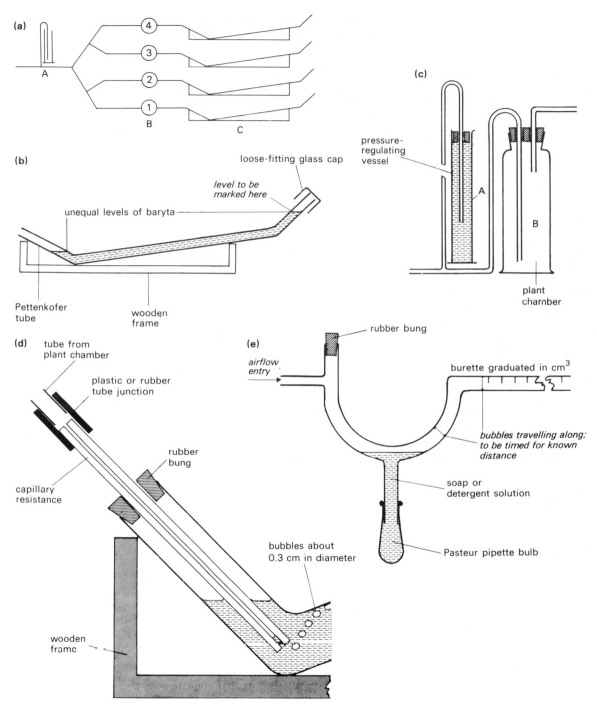

Figure B7.2 (a)–(d) Longitudinal section of improved Pettenkofer tube assembly for measuring respiration rates of plant material. (a) General layout of assembly; (b) levels of baryta water in Pettenkofer tube; (c) detail of pressure-regulating vessel and plant chamber; (d) detail of entrance to Pettenkofer tube with capillary resistance. (e) Longitudinal section of soap bubble gas flow meter.

A slide projector is recommended (see Part A, Expts 7.2 & 8.6) as the source of illumination.

blank chamber are required for gas-exchange calculations. Both chambers and the light source must be cooled by a fan to keep the temperature between 20 and 25 °C. Each chamber should contain 1–2 cm³ of water for uptake by the material. Because of its complexity the calculation of results is shown by an example.

SAMPLE RESULTS

Table B7.2 Titration of $Ba(OH)_2$ from Pettenkofer tubes with 1 mol m^{-3} (1 mM) HCl after a known volume flow of air has passed through the tubes.

Treatment	Volume flow of air (m³ h⁻¹ × 10⁻³)	Ba(OH)₂ volume in tube (m³ × 10⁻⁶)	Titration volume Start (m³ × 10⁻⁶)	Titration volume After 2 h (m³ × 10⁻⁶)	Leaf area (m² × 10⁻⁴)
(a) blank	2.5	105	2.8	1.3	—
(b) illuminated leaf	2.7	112	2.9	2.3	36
(c) darkened leaf	2.4	109	2.7	1.0	29

CALCULATION

(a) Carbon dioxide content of the air used in the experiment (given by results for blank)

$$= \begin{pmatrix} \text{difference} \\ \text{in titration} \\ \text{volume} \\ (\text{m}^3 \times 10^{-6}) \end{pmatrix} \times \begin{pmatrix} \text{relative} \\ \text{molar} \\ \text{concentration} \\ (\text{mol m}^{-3}) \end{pmatrix} \times \begin{pmatrix} \text{volume} \\ Ba(OH)_2 \\ \text{used} \\ (\text{m}^3 \times 10^{-6}) \end{pmatrix} \times \begin{pmatrix} [CO_2] \\ (\text{g mol}^{-1}) \end{pmatrix}$$

$$\div \begin{pmatrix} \text{time} \\ (\text{h}) \end{pmatrix} \div \begin{pmatrix} \text{flow} \\ \text{rate} \\ (\text{m}^3 \text{ h}^{-1} \times 10^{-3}) \end{pmatrix} \text{g } CO_2 \text{ m}^{-3}$$

$$= \frac{(2.8 - 1.3)10^{-6} \times 1 \times 105 \times 10^{-6} \times 44 \times 1 \times 1}{10^{-6} \times 2 \times 2 \times 2.5 \times 10^{-3}} \text{ g } CO_2 \text{ m}^{-3}$$

$$= 0.693 \text{ g } CO_2 \text{ m}^{-3}$$

(b) Amount of carbon dioxide supplied to illuminated chamber

$$= \begin{pmatrix} \text{flow rate} \\ (\text{m}^3 \text{ h}^{-1}) \times 10^{-3} \end{pmatrix} \times \begin{pmatrix} \text{time} \\ (\text{h}) \end{pmatrix} \times \begin{pmatrix} [CO_2] \\ (\text{g m}^{-3}) \end{pmatrix} \text{g } CO_2$$

$$= 2.7 \times 10^{-3} \times 2 \times 0.693 \text{ g } CO_2$$

$$= 3.742 \times 10^{-3} \text{ g } CO_2$$

Amount of carbon dioxide coming out of illuminated chamber (as equation under (a) above)

$$= \frac{(2.9 - 2.3)10^{-6} \times 1 \times 112 \times 10^{-6} \times 44}{10^{-6} \times 2} \text{ g } CO_2$$

$$= 1.478 \times 10^{-3} \text{ g } CO_2$$

Therefore, assimilation rate per unit leaf area for a leaf of $36 \times 10^4 \text{ m}^2$

$$= \frac{(3.742 - 1.478)10^{-3}}{2 \times 36 \times 10^{-4}} \text{ g } CO_2 \text{ m}^{-2} \text{ h}^{-1}$$

$$= 0.314 \text{ g } CO_2 \text{ m}^{-2} \text{ h}^{-1}$$

(c) Amount of carbon dioxide supplied to darkened chamber (as under (b) above)

$$= 2.4 \times 10^{-3} \times 2 \times 0.693 \text{ g } CO_2$$
$$= 3.326 \times 10^{-3} \text{ g } CO_2$$

Amount of carbon dioxide coming out of darkened chamber (as in equation under (a) above)

$$= \frac{(2.7 - 1.0)10^{-6} \times 1 \times 109 \times 10^{-6} \times 44}{10^{-6} \times 2} \text{ g } CO_2$$

$$= 4.076 \times 10^{-3} \text{ g } CO_2$$

Therefore, respiration rate per unit leaf area for a leaf of $29 \times 10^{-4} \text{ m}^2$

$$= \frac{(4.076 - 3.326)10^{-3}}{2 \times 29 \times 10^{-4}} \text{ g } CO_2 \text{ m}^{-2} \text{ h}^{-1}$$

$$= 0.129 \text{ g } CO_2 \text{ m}^{-2} \text{ h}^{-1}$$

and gross photosynthesis rate

$$= 0.314 + 0.129 \text{ g } CO_2 \text{ m}^{-2} \text{ h}^{-1}$$

$$= 0.443 \text{ g } CO_2 \text{ m}^{-2} \text{ h}^{-1}$$

The perforated zinc platforms conventionally used in reaction vessels should be replaced with very loose-fitting inverted plungers of plastic as shown in Figure B7.3 (zinc plus liquid, especially alkali, tends to release gas or heat).

The material should rest on a moist, not wet, disc of filter paper.

Manometers must be cleaned with chromic acid before filling with Brodie's solution (see Part C1) from a syringe. Beware of tiny bubbles of liquid in the supposedly empty portions of the manometer.

Both taps must be set open to the atmosphere when assembling the apparatus or taking it apart in order to avoid blowing out the manometer liquid.

After assembling the two reaction vessels, complete with taps open to the atmosphere, only *one* of the horizontal manometer limbs should be held in one hand and connected to the reaction vessel – otherwise breakages will occur. In the commercial model the taps and the manometer have solid, tapered fittings. If connections are made by tubing, plastic should be used, not rubber, but no play or bending must be allowed during measurements; the square end of the horizontal tap limb should be rounded off with a file.

If plastic 'Vygon' three-way taps (from Griffin & George Ltd (see Part C2)) are used, the lug preventing the tap from turning through 360° should be cut off with a sharp knife and the tap set only in either of two ways: ⊢ open to the atmosphere, or T connecting the two reaction vessels via the manometer. The tap should be fitted into the rubber bung directly, not via a metal tube, which makes a slippery joint.

7.3   Respiration rate and respiratory quotients

By Professor H. Meidner, University of Stirling
cf. Expts 11.6 & 11.7

The Barcroft apparatus shown in Figure B7.3 can be made from laboratory glassware. A neat, commercially available version (from Griffin & George Ltd, see Part C2) requires minor modifications.

Seeds at different stages of germination or of different kinds are the most suitable material, but flower buds, storage tissue, darkened leaves or fruit can be used.

The first set of measurements is taken with 5 cm³ water in both reaction vessels and the three-way stopcock closed to the atmosphere. If the volume of oxygen taken in by the material in the reaction vessel equals the volume of carbon dioxide given off (case 1), the manometer will remain at its zero reading. This will be apparent within 5 min. After setting the three-way taps open to the atmosphere, a second reading is taken and then a third with the stopcock closed to the atmosphere. If the manometer liquid is depressed on the side of the reaction vessel containing the material, more carbon dioxide is being produced than oxygen taken in (case 2). If the reverse gas exchange has occurred, the manometer liquid will rise on the side of the vessel containing the material (case 3).

The second set of measurements is taken with the same material in the system but with 5 cm³ of 4×10^3 mol m^{-3} (4.0 M) NaOH in both reaction vessels. Carbon dioxide is thus taken out of the gas volume and the timed manometer reading is the absolute measure of the rate of oxygen intake by the material. A difference of about 4 scale units between the manometer limbs should be reached before the taps are set open to the atmos-

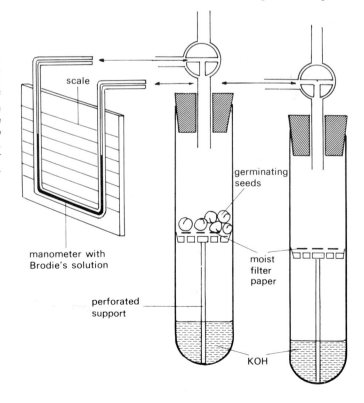

Figure B7.3 Longitudinal section of Barcroft reaction vessels and the manometer in perspective to show how it fits onto the three-way stopcocks and can hang over the rim of the water-bath in which the reaction vessels are positioned.

Before placing the reaction vessels in a beaker with water at room temperature, or in a constant temperature water-bath, the system should be tested for airtightness by placing one hand around one of the reaction vessels while taps are set to connect the two reaction vessels. The manometer liquid should move immediately.

The reaction vessels must be immersed in the water-bath right up to the base of the rubber bung. There must be no exposed airspace in the vessels above the water level.

The 4×10^3 mol m^{-3} (4.0 M) NaOH solution should have dissipated its heat of solution before being used, and the greatest care must be taken to avoid contact with the live material.

As the reaction is fast, everything except the addition of NADH (DPNH) must be completed beforehand, including the fitting of the sealing cap to the electrode; only then should the electron donor be added.

phere and second and third measurements are taken in order to obtain a mean value.

For the calculation of the respiratory quotient (RQ), the denominator is the absolute volume of oxygen taken in when NaOH is used in the system, for instance 36 vol. units h^{-1}. From the measurements with water in the reaction vessels in case 1 (manometer remains at zero), the volume of oxygen = the volume of carbon dioxide; hence the RQ is 36/36 = 1.0, indicating carbohydrate as the respiratory substrate. For case 2 (manometer liquid depressed on side of reaction vessel with material, i.e. more carbon dioxide produced than oxygen taken in) the RQ will be $(36 + x)/36$, i.e. >1.0, indicating partial anaerobiosis. In case 3 with less carbon dioxide given off than oxygen taken in, the RQ will be $(36 - y)/36$, i.e. <1.0, indicating that lipids or proteins have been respired. x and y are the volume units per hour calculated from the measurements with NaOH in the reaction vessels.

7.4 Respiratory electron transport and oxidative phosphorylation
Based on information given by Dr J. M. Palmer, Imperial College of Science and Technology
cf. Expts 4.3, 8.5 & 8.6

The experiment described here measures the O$_2$ consumption by isolated mitochondria (see Part A) from a medium containing different amounts of ADP.

The oxygen electrode to be used as the measuring instrument must be calibrated as in Experiment 8.6 with an air-saturated medium for full-scale deflection and a medium deoxygenated by Na$_2$S$_2$O$_4$ for zero setting. At 25 °C oxygen-saturated water contains 240 nmoles O$_2$ cm^{-3}. Oxygen consumption is measured by using 1.8 cm^3 of the reaction medium in the electrode vessel for a 1.6-cm-diameter electrode, and less for smaller electrodes. The *sensitivity* (not the zero setting) must be adjusted to give near full-scale deflection on the recorder and the system allowed to run for 1 min so that a reasonably steady tracing is obtained. Then 0.1 cm^3 of the mitochondrial suspension is added which will probably cause a downward deflection of 5–10%, and the electrode sealed. After a further minute the electron donor, 0.1 cm^3 of 15 mol m^{-3} (15 mM) NADH (DPNH), is added via the fine bore of the plug, ensuring that the liquid just enters the fine bore.

After 1–1½ min, while O$_2$ is being consumed and the recorder has reached 75–80% of full-scale deflection, 10 mm^3 of 25 mol m^{-3} (25 mM) ADP are injected through the fine bore in the plug. The new rate of O$_2$ consumption will now be recorded and after an increase it will gradually return to its original rate. Further additions of 5, 15 and 25 mm^3 of 25 mol m^{-3} ADP will give a set of four ADP concentrations from which can be calculated the rate of O$_2$ consumption in nmoles per milligramme of protein per minute without ADP, a maximum rate with ADP present and a final rate after the exhaustion of ADP. The graph of O$_2$ consumption against concentration of ADP represents the rate of O$_2$ consumption before it has returned to its initial slow rate. The slope of this line gives the ADP/O$_2$ ratio, which is a measure of the number of sites of oxidative phosphorylation (external NADH = 2.0).

8 Photosynthesis

8.1 Infra-red gas analyser experiments
Based on information given by Dr A. Goldsworthy, Imperial College of Science and Technology
cf. Expts 7.2, 8.2 & 8.3

(A) RATES OF PHOTOSYNTHESIS

The use of calibrated gas analysers provides excellent opportunities for the study of rates of photosynthesis of reductive pentose phosphate pathway (C_3) and 4-C dicarboxylic acid pathway (C_4) plants and is most readily achieved when the analyser is used in a differential open circuit. For such measurements accurate flow rates must be known as well as the carbon dioxide concentration of the incoming air. Measurements can include studies of temperature effects on rates of photosynthesis.

Alternatively, a closed circuit can be used with repeated small injections of known volumes of carbon dioxide and recording the slopes of the assimilation curves at known $[CO_2]$.

(B) RATES OF RESPIRATION

Measurements of respiration in the dark are the easiest to accomplish in either the open or the closed circuit as the effects of ambient concentrations of carbon dioxide (within reason) on respiration in the dark are very small. An open circuit will be most sensitive if CO_2-free air is supplied instead of 'atmospheric' air.

(C) ESTIMATES OF PHOTORESPIRATION

Using C_3 and C_4 species, the rate of carbon dioxide release into CO_2-free air from illuminated material and the *post-illumination burst* should be observed. This is best done in an open circuit system. The recorder will show a steady rate of carbon dioxide evolution in the light due to photorespiration. This represents, however, an underestimate of the true rate due to refixation of some of the carbon dioxide evolved. Varying flow rates should be tried to emphasise the point. When the lights are switched off, refixation of carbon dioxide stops instantaneously, but photorespiration continues for about 1 min. This is seen as a temporary increase in carbon dioxide evolution – the post-illumination burst. The magnitude of this burst gives a better estimate of the rate of photorespiration. After the initial burst the rate of carbon dioxide evolution gradually declines to the rate of dark respiration at the prevailing temperature. The post-illumination burst should be observed after different photon-flux densities as its magnitude depends on the preceding level of illumination.

(D) CARBON DIOXIDE COMPENSATION POINTS

These must be measured in closed circuits and greatest differences will be measured between C_3 and C_4 species. The effect of temperature on C_3 and C_4 compensation points is also different. Carbon dioxide compensation points should be approached both from 'below', starting with CO_2-free air in the closed system, and from 'above', beginning with ambient air in the system.

8.2 Simplified infra-red gas analyser system for carbon dioxide compensation points
Based on information given by Dr A. Goldsworthy, Imperial College of Science and Technology
cf. Expt 8.1

Mylar bags, 14 × 45 cm, practically impermeable to CO_2, are obtainable from E. I. du Pont de Nemours & Co. (see Part C2).

Portions of leaves or whole leaves with their lower surface upwards are floated on 15 cm³ distilled water in Petri dishes and placed inside Mylar bags lying horizontally on the bench. One bag should be inflated with ordinary air from a pump and a parallel sample flushed repeatedly with CO_2-free air. When inflated, the bags can be closed by twisting *clockwise* and fastening the twist with a clothes peg.

After illumination at a reasonably constant known temperature at 50–100 μmol m⁻² s⁻¹ (50–100 μE m⁻² s⁻¹) for about 1 h, a 5 cm³ graduated pipette is connected at its mouthpiece to a small magnesium perchlorate drying tube and hence directly to the gas analyser. The tip of the pipette is then inserted into the Mylar bag up to the peg and the bag twisted tightly around the pipette in an *anticlockwise* direction *and held in place by hand*. After removing the clothes peg, manual pressure applied to the bag will unroll the clockwise twist and the air can be expelled from the bag into the gas analyser for a reading of the [CO_2]. The analyser system should not be in excess of about 200 cm³ so that it can be completely flushed by the contents of the bag.

As the one sample establishes the compensation point from zero [CO_2] and the other from atmospheric [CO_2], a good check on the reliability of the method is obtained automatically; a discrepancy usually indicates that insufficient time has been allowed for the compensation point to have been reached. Many samples can be analysed at a rate of about one per minute.

8.3 Light compensation points
By Professor H. Meidner, University of Stirling
cf. Expt 8.1

The light compensation point is the photon-flux density at which an illuminated leaf maintains the normal carbon dioxide concentration of the atmosphere, i.e. 600–640 mg m⁻³ (300–320 p.p.m.). Light compensa-

For the measurement of photon-flux density, the PAR sensor should be inside the leaf chamber and in such a position that it measures the irradiance at the leaf surface. Leaf temperature should be measured with a thermocouple (see Part A) or thermistor attached to the leaf.

Shade leaves may have light compensation points at between 20 and 50 μmol m^{-2} s^{-1} (20–50 μE m^{-2} s^{-1}) and sun leaves at between 50 and 200 μmol m^{-2} s^{-1} (50–200 μE m^{-2} s^{-1}).

Measurements of photon-flux density should also be made with the different leaves lying on the sensor to ascertain the percentage of light absorbed and transmitted. This should be related to measurements of leaf thickness.

tion points differ between leaves of different species and between sun and shade leaves of the same species; they increase with leaf temperature. Light compensation points are most readily measured in infra-red gas analyser circuits in which irradiation, measured as photosynthetic active radiation (PAR) (μmol m^{-2} s^{-1}), can be adjusted by raising or lowering the light source.

Suitable materials are sun and shade leaves of beech (*Fagus sylvatica*) or copper beech in which the two are differently pigmented, and of balsam (*Impatiens parviflora*). For a fibrous leaf *Rhododendron* spp. can be used and a comparison can be made between a C_4-leaf like maize (*Zea mays*) and any of the C_3 cereal leaves.

It will be found that shade leaves have distinctly lower light compensation points than sun leaves and it is the latter that show the greatest change in light compensation with temperature.

REFERENCE

Meidner, H. 1970. *Nature* **228**, 1349.

RELEVANT ANATOMICAL STUDIES

Leaf thickness

Using new, sharp razor blades, transverse, hand-cut sections of leaf blades can be cut between pieces of elder pith or expanded polystyrene. Total leaf thickness, and epidermal and mesophyll tissue thicknesses should be measured using an eyepiece graticule (see Part A). If possible, the palisade and spongy mesophyll tissues should be measured separately. These measurements are especially relevant to sun and shade leaves of the same species, or for comparisons between sun and shade leaves generally. The measurements also have a bearing on the percentage of light transmission of these leaves.

When dealing with sun and shade leaves, their respective epidermal (and other) average cell sizes in surface view and their stomatal frequencies are of interest. These properties vary also with height of insertion; it is therefore necessary to make comparisons of leaves of comparable positions on the plants.

8.4 Carbon dioxide compensation points without infra-red gas analyser

By Professor H. Meidner, University of Stirling
cf. Expts 7.2, 8.1 & 8.2

Baryta water and CO_2-free water must be kept away from the open air, i.e. in completely filled stoppered bottles. Syringes filled with the liquids are a good means of storing these during classes.

Ten cubic centimetres of 5 mol m^{-3} (5 mM) baryta water are recommended, as the amount of carbon dioxide at the CO_2-compensation point of a C_3 leaf kept in a 1000 cm^3 container will be no more than about 0.1 mg.

If infra-red gas analyser (Expts 8.1–8.3) work is not possible, it is recommended to enclose leaves of comparable area in 1000-cm^3 measuring cylinders closed with Parafilm or Mylar bag material (cf. Expt 8.2) and illuminated for about 50 min. Ten cubic centimetres of clear baryta water (see Part C1) are then injected from a syringe, the pierced hole is closed with Parafilm and the vessel left illuminated for a further 20 min with occasional swirling of the baryta water around the base of the cylinder. Visually, 4-C dicarboxylic acid pathway (C_4) leaves will have clear liquid at their base, and reductive pentose phosphate pathway (C_3) leaves

Transfer of liquids, titrations and additions to indicator must be carried out quickly.

slightly turbid liquid. Quick transfer of the baryta water into a small, capped conical flask and titration with 1 mol m^{-3} (1 mM) HCl will demonstrate quantitatively the near zero and about 100 mg m^{-3} (50 p.p.m.) CO_2-compensation points. A blank and, if desired, a 'dark' treatment should be run.

Another procedure involves the use of universal indicator. The leaves should be kept with their base, petiole or small stem portion, in 1 cm^3 water in a small beaker at the bottom of the measuring cylinder. After 40 min illumination 10 cm^3 CO_2-free water are injected and the vessel is illuminated for another 30 min with occasional swirling of the liquid. After 30 min most of the water is withdrawn with a pipette and delivered into a small beaker containing 1 cm^3 indicator.

It is also possible to set up the leaves with their base or petiole submerged in water in small beakers hanging by a thread from a neoprene (not rubber) bung inside a 250 cm^3 conical flask. The flask must be sealed by Parafilm or Mylar (from E. I. du Pont de Nemours & Co., see Part C2) material over the bung. Ten cubic centimetres of a solution of 3 mol m^{-3} (3 mM) $KHCO_3$ plus 0.1 cm^3 universal indicator are added to the flask at the beginning of the experiment and the resultant colour change with C$_3$ and C$_4$ leaves is observed after 1 h illumination. Before starting the experiment with leaves CO_2-free air should be passed through the $KHCO_3$ solution until no further colour change occurs.

Water from C$_4$ leaves will be green, and that from C$_3$ leaves yellowish green. Blank (and dark treatments) should be added.

RELEVANT ANATOMICAL STUDIES

A comparison of transverse, hand-cut sections of maize leaves and of a C$_3$ cereal leaf will highlight the features of the *Kranz* anatomy. Measurements of mesophyll cell diameters and chloroplast dimensions should be carried out (cf. isolated chloroplasts Part A). If it is possible, protoplast suspensions of the two kinds of mesophyll should be prepared to show the large bundle sheath cells with their chloroplasts in maize.

8.5 Photosynthetic potential of plant parts
Based on information given by Dr R. O. Mackender, Queen's University, Belfast cf. Expt 7.4

Phaseolus vulgaris leaves or leaves with epidermal trichomes are unsuitable since they tend to 'clump' together.

For measuring photosynthetic potentials of plant parts by means of an oxygen electrode, seedlings of either oats (*Avena sativa*) with 9- to 10-cm-long first leaves or pea (*Pisum sativum*) are recommended. The pea seedlings can be divided into root and shoot sections, and the latter, if desired, into apex, individual leaves numbered from the base, stipules and stem sections. From the oat seedlings about 20 first leaves should be neatly cut into 9 × 1 cm long segments numbered from the leaf base.

The oxygen electrode should be set so that the meter reads 0.4 (40% of full scale) in the presence of 5 cm^3 of 50 mol m^{-3} (50 mM) HEPES–NaOH buffer (see Buffers, 'Good', Part C1) at pH 7.6 containing 20 mol m^{-3} (20 mM) $NaHCO_3$. Four square centimetres of each tissue type are needed for each measurement and this must be *sliced* into 1-mm-wide strips with a *new* razor blade by placing the tissue in a Petri dish containing 10 cm^3 of about 0.5 mol m^{-3} (0.5 mM) $CaSO_4$ so that the tissue and the cut surfaces remain at all times submerged in the $CaSO_4$ solution. The sliced tissue (originally 4 cm^2) is then transferred into the oxygen electrode for measurements in the dark (covered with black cloth) and in the light (provided by a slide projector).

Instructions to students must emphasise that chopping the tissue renders it useless. The tissue must be *'sliced'* with a brand-new razor blade.

About 0.5 mol m^{-3} (0.5 mM) $CaSO_4$ implies a saturated solution.

Results can be expressed on a weight basis if green and non-green tissue are used to obtain an energy budget. Comparable measurements using green tissue only should be on a chlorophyll basis or area basis. Net-photosynthesis is expressed in mm^3 or μmol O_2 per unit chlorophyll per unit time, and similar units on an area basis are chosen for respiration measurements.

REFERENCE

Jones, H. G. and C. B. Osmond 1973. *Aust. J. Biol. Sci.* **26**, 15–24.

RELEVANT ANATOMICAL STUDIES

LEAF AIRSPACES

Vicia faba leaves are recommended for these observations. The lower epidermis should be peeled off (see Part A) and discarded so that the exposed spongy mesophyll tissue can be examined under the microscope without a coverslip and with a fairly strong illumination from below and good vertical focusing to different depths.

To show the hydrophobic nature of the mesophyll cell walls and the effect of the high surface tension of water (see Expt 1.6), a small droplet of water should be placed on the exposed mesophyll tissue where it will remain as a spherical droplet. On being touched with a needle point that has been dipped into a surfactant, it will immediately spread, infiltrating the airspaces.

The upper epidermis is not as readily peeled off as the lower, but with a pair of pointed forceps enough of the upper epidermis can be stripped off to reveal the underlying palisade tissue of cells appearing round in surface view, with smaller but not inconsiderable airspaces between them.

DEVELOPMENT OF MESOPHYLL TISSUE

Recommended for these observations are tobacco leaves (*Nicotiana tabacum*) of different ages: very small, about 1 cm long; somewhat more developed leaves about 3 cm long; and mature leaves. No. 6 cork-borer discs (0.9 cm internal diameter) should be cut out of the laminae and separately infiltrated with a solution of 3 g chromium tri-oxide (CrO_3) in 100 cm³ water by placing 10 cm³ of the solution in a 50 cm³ beaker containing the leaf discs and exposing the discs repeatedly (10 times) to partial vacuum and release of vacuum in a vacuum desiccator. The tissue is left overnight in the solution or longer, until, after decanting the acid, a touch with a rounded glass rod causes the tissue to disintegrate. Instead of decanting, a Pasteur pipette can be used. After the tissue has been found to disintegrate, it is ground with a rounded glass rod in the trace of liquid left, until a dense suspension is obtained. This must now be washed 3 times with 10–20 cm³ water, allowed to settle or centrifuged and the water discarded each time. The final cell mass should be suspended in about 3 cm³ water for use in the laboratory. A drop of the suspension observed under a coverslip at 100× or 400× magnification will show the jigsaw-puzzle shaped cells of the spongy mesophyll, the smaller rectangular-shaped cells of the palisade layer, glandular hairs and some epidermal cells. Frequencies of the different mesophyll cell types in the preparations made from leaves of different ages, and chloroplast counts in the two types of cells, should be recorded.

8.6 Photosynthetic action spectra
**Based on information given by Dr K. Hardwick, University of Liverpool
cf. Expt 7.4**

A cause of failure can be the saturation of the suspension medium with oxygen; this is indicated when the oxygen concentration does not rise, or ceases to rise on illumination. The remedy for this situation is to remove the electrode stopper, allowing for equilibration with the atmosphere. The oxygen concentration will be observed to decrease rapidly. To prevent the suspension medium becoming saturated with oxygen, it is best to keep the stock culture in the dark during the practical class and also to darken the electrode except when readings are being taken. These should not involve too lengthy exposures to light.

Extraneous light must be excluded from the electrode during *all* measurements by providing a black cloth hood for the electrode. In order to admit a light beam, a slit should be provided that can be closed for measurements of dark respiration. Diffuse laboratory light may be significant in causing photosynthesis when the illuminating wavelength band is only weakly effective for photosynthesis.

A better measure of light is probably obtained by positioning the meter sensor behind half of an old electrode obtained by sawing in half longitudinally through the centre of the sample chamber. The photon-flux measurements with the different filters should be made in one session once the sensor has been positioned. This will save time and avoid the repeated placing of the sensor. Measurements with the eight filters can be carried out in a few minutes.

Light supply *must* be measured with a PAR sensor; if a W m^{-2} sensor is used it must be cross calibrated with a PAR sensor in μmol m^{-2} s^{-1} (μE m^{-2} s^{-1}) (this in itself is instructive). Either a Licor-type meter from Licor Instruments (see Part C2), or best a spectroradiometer, is needed. Quantum flux measurements should be made before each new filter is used, and preferably the same number of μmol m^{-2} s^{-1} used with each filter. Otherwise, unless the quantum flux used is below half the saturating light supply for photosynthesis of the algal suspension, serious errors will arise, even when corrections are attempted for differences in quantum flux.

The experiments described here involve the use of the oxygen electrode as a means of measuring O_2 exchange during respiration in the dark and photosynthesis. The use of the oxygen electrode must be fully understood, but the scope of this book does not permit discussion of it here.

For action spectra six to eight good colour filters covering the range 400–700 nm (4000–7000 Å) should be used together with a suitable light source (see below). Most suitable materials are suspensions of *Scenedesmus* or *Chlorella* cultured as described in Part A. As a preparatory procedure, after calibrating the electrode, it is recommended to put some algae into the cuvette and to measure O_2 evolution under full illumination and O_2 consumption with the electrode darkened. Photosynthesis should be at least twice the rate of respiration. This will confirm the satisfactory working of the electrode and indicate the speed with which the material responds to light-on/light-off treatments.

When ready for measurements using different wavelengths, the algae should be allowed to respire briefly in darkness until the O_2 concentration decreases to about the original level prior to illumination before the next filter is used. These dark measurements together will give a mean value for respiration rate (cf. Expt 7.4). All measurements should be made from approximately the same oxygen concentration at the start and between comparable O_2 and CO_2 concentrations.

Results are expressed as rates of O_2 evolution in μmol O_2 cm^{-3} min^{-1} at photon-flux densities of the measured photosynthetic active radiation (PAR) in μmol m^{-2} s^{-1} (μE m^{-2} s^{-1}). The same PAR must be used for each filter so that photosynthetic rate can be plotted against wavelength to produce a graphical action spectrum plot.

COLOUR FILTERS

Reasonably priced filters (£40.00 for a set of eight) can be prepared by taking the filters of a Corning 252 colorimeter (see Part C2) out of the frames in which they are mounted. Filters can then be mounted, between glass if preferred, in 35-mm slide-holders. Peak transmissions are at 430, 470, 490, 520, 540, 580, 600 and 710 nm (4300, 4700, 4900, 5200, 5400, 5800, 6000 and 7100 Å). These filters must not be left illuminated in the projector for longer than essential – otherwise they become opaque. Other suitable, but much more expensive, filters are Balzer (see Part C2) with 30 nm (300 Å) half-band widths.

LIGHT SOURCE

Quartz iodide 150 W bulbs in a slide projector should give about 50 μmol m^{-2} s^{-1} at 430 nm (4300 Å); this photon-flux density can be obtained at the other wavelengths by increasing the distance between electrode and projector. However, the light beam must be adjusted accurately in respect of the centre of the electrode so that the illumination of the material is the same for all settings. The beam should point slightly downwards to avoid shading by the electrode locking ring. The projector positions are best marked on the bench with chalk or tape. For the

The comparatively narrow band filters will allow for only low rates of photosynthesis; in many cases these will appear on the recorder trace only as a decreased rate of oxygen uptake, i.e. the downward slope of the oxygen uptake line in the dark will be less steep in the light. Thus measurements of net rates of photosynthesis depend heavily on reliable values for dark respiration.

The alternative way of presenting a plot of action spectra is to plot the reciprocal of the PAR at each wavelength required to produce a fixed amount of oxygen, but with the equipment used here this is much more difficult to obtain.

A further useful exercise to help familiarise the students with the equipment is the measurement of net-oxygen exchange of the algal suspension at a range of light intensities, so as to construct a curve showing net-photosynthesis rate *v.* light intensity response. Measurements at individual light intensities need only be continued for as long as is needed to obtain a linear response on the chart recorder – often only 1 or 2 min.

measurement of PAR it is convenient to place some plasticine on the magnetic stirrer top in place of the electrode and fix the quantum sensor accurately in the position corresponding to the centre of the electrode cuvette.

ADDENDUM
(By Dr J. Hannay, Imperial College of Science and Technology.)

For the estimation of quantum efficiency at different wavelengths, volume and surface area of the algae should be calculated so that μmol O_2 s^{-1} and photon-flux density in μmol s^{-1} can be computed. Two estimates are required:

(1) It can be assumed that the area on which incident PAR falls is the radial longitudinal section (RLS) of the algal suspension in the oxygen electrode. The RLS can be measured from d, the diameter of the inner vessel of the electrode, and H, the height of the algal suspension in that vessel. If H is not readily measured, it can be calculated as follows: known volume of suspension = $\pi (d/2)^2 H$ cm³ therefore,

$$H = \frac{\text{volume}}{\left(\pi \dfrac{d}{2}\right)^2} \text{ cm}$$

(2) The per cent transmission of a sample of the same algal suspension is measured in a spectrophotometer at the same wavelengths as are used with the interference filters during the experiment. Thus, if there are, for instance, 100 units cm^{-2} s^{-1} incident radiation and there is found to be 20% transmission, 80 units cm^{-2} s^{-1} must have been absorbed. If the RLS of the algal suspension has been calculated to be, for instance, 3.2 cm², then 80×3.2 μmol s^{-1} are absorbed by the algal suspension (assumed to be by the algal cells alone).

CALCULATIONS

$$\text{Quantum yield} = \frac{\mu\text{mol } O_2 \text{ s}^{-1}}{\mu\text{mol photons s}^{-1}}$$

Since the energy content per photon = hc/λ the energy content of 1 mole photons (E)

$$E = \frac{N h c}{\lambda} = \frac{1.2 \times 10^8}{\lambda} \text{ J mol}^{-1} \text{ photons}$$

and that of 1 μmol photons $= \dfrac{1.2 \times 10^2}{\lambda} \text{ J}$

where N is Avogadro's number = 6.02×10^{23} mol^{-1}, h is Planck's constant = 6.63×10^{-34} J s, c is the speed of light = 3×10^8 m s^{-1}, λ is the wavelength of light used, measured in nm, thus the energy absorbed $= \dfrac{\mu \text{mol photons} \times 1.2 \times 10^2}{\lambda}$ J

$$= \frac{120}{\lambda} \text{ J } \mu\text{mol}^{-1}$$

The estimation of the per cent energy conversion efficiency of light into carbohydrate is based on the assumption that for every mole of O_2 evolved 1 mole of CH_2O is synthesised; this is equivalent to the incorporation of 470 kJ mol^{-1} O_2 evolved so that

$$1 \ \mu\text{mol } O_2 \equiv 0.47 \text{ J}$$

therefore the per cent efficiency

$$= \frac{\mu\text{mol } O_2 \text{ evolved s}^{-1} \times 0.47}{\mu\text{mol photons absorbed s}^{-1} \times 120/\lambda} \times 100$$

8.7   Measurement of rates of photosynthesis
Based on information given by Professor D. A. Baker, Wye College cf. Expt 7.2

Experiments involving the use of radioactive material must be carried out according to the safety regulations in force. All radioactive waste, including plant material, must be disposed of immediately after the conclusion of experiments.

The experimental procedure outlined below serves to compare rates of photosynthesis between leaves of different heights of insertion and age and from different species or under different light treatments. As comparisons must be made in relation to leaf area, leaf discs cut with cork-borer No. 6 (0.9 cm internal diameter) are used. Maize (*Zea mays*), sunflower (*Helianthus annuus*), runner bean (*Phaseolus vulgaris*) and broad bean (*Vicia faba*) leaves are recommended.

Transparent plastic Petri dishes, each with a moist filter paper and a centrally placed aluminium planchette, are used as treatment chambers. A hole the size of a syringe needle must be drilled, or more easily burnt, into the centre of the lid. Ten discs, all with the same epidermis upwards, are placed around the central planchette at a distance of about 2 cm from it and a droplet containing 5 μCi NaH^{14}CO$_3$ solution is placed in the planchette. When the lid has been closed and sealed, a drop of 10^3 mol m^{-3} (1.0 M) HCl is injected so that ^{14}CO$_2$ will be released; after this the hole must be sealed as well. The illuminated dishes should be kept for 10 min in a fume cupboard.

The volume of NaH^{14}CO$_3$ depends on the specific activity of the solution and the volume of 10^3 mol m^{-3}. HCl must be in excess of the total NaH^{14}CO$_3$ so that all ^{14}CO$_2$ is liberated.

After 10 min the leaf discs are removed with forceps and three discs are glued into each marked planchette for drying at 70 °C for 30 min. One-minute counts (minus background) are made to obtain mean counts min^{-1} per disc for the different treatments. In order to express the results on a unit fresh and dry weight basis, parallel samples for these weight determinations should be used.

8.8   Assay of chloroplast activity – Hill reaction
Based on information given by Dr R. Phillips, University of Stirling

The assay medium is made up from:

15 cm^3	10^3 mol m^{-3} sorbitol (1.0 M)
12 cm^3	75 mol m^{-3} tricine (75 mM)
2.4 cm^3	25 mol m^{-3} NaEDTA (25 mM), pH adjusted to 7.6 with either KOH or HCl
12.0 cm^3	10 mol m^{-3} NaHCO$_3$ (10 mM)
18.6 cm^3	distilled water

If cuvettes are not available and test-tubes are used, each tube should be marked with a vertical line, 1 cm long, near its mouth so that it can be placed every time in the same position in the spectrophotometer holder.

In order to measure the absorbance of the suspension, 2.8 cm^3 of this medium plus 0.1 cm^3 of a chloroplast suspension (see Part A) are put into the cuvette of a spectrophotometer at 600 nm (6000 Å). Thereafter, 0.1 cm^3 of 1 mol m^3 dichlorophenol-indophenol dye (DCPIP) is added, the cuvette inverted once for mixing and the absorbance determined immediately in the dark, representing the starting point. This measurement should be completed during a standard time, e.g. 7 s. Next, the cuvette is exposed to a light source (see Expt 8.6) outside the spectrophotometer for exactly 20 s. The contents are mixed by inverting the tube once, and the absorbance is measured during the standard length of time as for the starting value. This procedure is repeated with exact time-keeping and standard mixing procedure by one inversion of the cuvette until the absorbance no longer changes. (It may not return to the starting value.)

If the decline in absorbance is too fast for repeated measurements, the chloroplast suspension must be diluted or the volume of the DCPIP dye increased.

The increase in absorbance when dye is added is equivalent to a known amount of dye and the results are plotted as µmol dye on the y-axis and time on the x-axis. The initial rate when there is a linear decline in absorbance can be calculated. Using the data for chlorophyll content of the reaction mixture, Hill activity can be expressed as µmol DCPIP reduced min^{-1} unit chlorophyll^{-1}.

REFERENCE

Trebst, A. 1972. *Methods in enzymology*, vol. 24b, 146–65. New York: Academic Press.

RELEVANT ANATOMICAL STUDIES

Chloroplasts in C$_3$ and C$_4$ mesophyll cells and those in isolated protoplasts (see Part A) of C$_4$ bundle sheath cells should be compared with 'isolated chloroplasts' in the same suspensions as are used for studies of electron transport capacities of such chloroplasts. Observations under oil immersion will show swollen grana of thylakoids rather than intact chloroplasts, as the stroma and outer membrane will have been lost in the preparation of the suspension. There will be some contaminating debris and possibly some bacteria. Dimensions of the different structures should be measured.

8.9   Chlorophyll content of chloroplasts

METHOD 1
(Based on information given by Dr R. Phillips, University of Stirling.)

A chloroplast suspension is prepared (see Part A) and 0.1 cm^3 is added to 2 cm^3 of 80 cm^3 acetone diluted with 20 cm^3 water in a spectrophoto-meter tube. It is shaken and allowed to stand for 5 min; then absorbance is measured at 645 and 663 nm (6450 and 6630 Å). A tube containing diluted acetone is used to zero the instrument. The reading should be less than 0.8; if it is greater, accurate dilution with diluted acetone is required. Both wavelengths must be measured at the same dilution. Using the above measurements,

$$\text{chlorophyll concentration} = 20.2\, A_{645} + 8.02\, A_{663}\ \mu\text{g cm}^{-3}$$

where A is absorbance at the wavelength indicated.

REFERENCE

Arnon, D. I. 1949. *Plant Physiol.* **24**, 1–15.

METHOD 2
(Based on information given by Professor D. A. Baker, Wye College.)

An extract of chlorophyll in petroleum ether is prepared and 1 cm^3 is diluted with 9 cm^3 acetone for measurements of absorbance at 645 and 663 nm (6450 and 6630 Å); zero setting is obtained with acetone. If necessary, dilution by a known factor must be undertaken.

By using the nomogram in Figure B8.9, spectrophotometer readings can be transformed into chlorophyll concentrations. By drawing a line connecting the readings at 663 nm (6630 Å) (column 2) and those at 645 nm (6450 Å) (column 4), total chlorophyll concentration $(a + b)$ can be read off where the line cuts column 3. Chlorophyll a concentration is given where the line cuts column 1 and chlorophyll b concentration where it cuts column 5. Ratios of a:b can now be calculated for the extracts from different species.

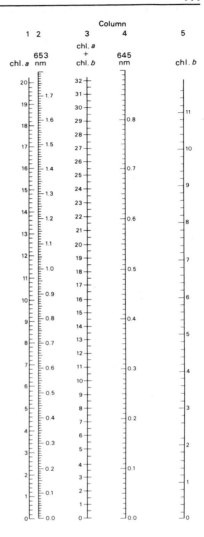

Figure B8.9 Nomogram allowing for the estimation of chlorophyll a and b concentrations from their absorbance at 645 and 663 nm (6450 and 6630 Å). (Based on Kirk, J. T. C. 1968. *Planta* **78**, 200.)

8.10   Properties of photosynthetic pigments
Based on information given by Dr D. N. Price, Plymouth Polytechnic; and Dr K. Hardwick, University of Liverpool

(A)   EXTRACTION AND SEPARATION

METHOD 1
A wide range of materials can be used, including sun and shade leaves as well as leaves from reductive pentose phosphate pathway (C_3) and 4-C dicarboxylic acid pathway (C_4) plants, and fronds of green, brown, red and blue-green algae (*Ulva lactuca*, *Fucus spiralis*, *Palmaria palmata*, *Nostoc* or *Anabaena* spp).

About 2 g of the material are thoroughly ground in a mortar with some sand and about 15 cm^3 acetone. For the brown algae 3 g and for the red 4 g are recommended. After filtering the extract through Whatman No. 1

Commercial frozen peas are also suitable material from which to prepare acetone extracts.

Deep frozen marine algae can be used in this experiment if fresh material is difficult to obtain. The red phycobilin pigments are not extracted and a reddish purple residue remains in the red algal extract.

CAUTION: Chloroform and acetone can form an explosive mixture. Although in this procedure they do not come into contact, it is important to avoid contact while handling and on no account must these liquids be mixed in waste bottles. Diethyl ether is flammable!

The loading of the extracts on the plates should be 'light' and only slightly 'overloaded' if the lower levels of some of the carotenoids are to be identified by spectrophotometry.

Satisfactory separations of pigment extracts can be achieved at less cost using thin-layer 'plates' as small as microscope slides, the separation being achieved with the same solvent mixture in 20 min. It is, of course, difficult to separate sufficient of the individual pigments on this scale for spectrophotometry, but the demonstration of the separation is good. The flexible, plastic-backed plates can be cut with scissors to give many small ones the size of microscope slides.

paper into a 100 cm³ separating funnel, about 10 cm³ diethyl ether are slowly added while swirling the liquids in the funnel, followed by the addition of about 10 cm³ distilled water and gentle agitation to avoid the formation of emulsions. Acetone, ether and distilled water should be in proportions $3:2:2$. The liquids are next allowed to separate for 2 min after which the lower acetone–water mixture is run off and discarded. Another washing with 10 cm³ distilled water is next carried out before the remaining pigment extract is transferred to a 100 cm³ stoppered flask containing about 10 g of freshly dried anhydrous Na_2SO_4. The flask is now allowed to stand stoppered in the dark and after 1 h the dried extract is decanted and evaporated to dryness in a rotary evaporator. The residue must be dissolved in a minimum volume of chloroform and placed in a vial suitable for volume reduction by connecting it to a vacuum pump. A very concentrated extract is obtained for loading on to thin-layer chromatographic plates coated with a 250 μm layer of silica gel G.

The plates are run in darkness with a solvent of hexane–diethyl ether–acetone ($60:30:20$) until the solvent front reaches the top 3 cm region of the plate; it must *not* be allowed to reach the top edge of the plate. The appearance of the plates will be similar to that shown in Figure B8.10. Pigments can be identified by their positions on the plate and this can be confirmed by spectrophotometry of eluted bands in standard solvent systems.

METHOD 2
(By Professor D. A. Baker, Wye College.)

About 1 g of fresh leaf tissue is cut into narrow pieces and ground in a mortar with some sand. The ground-up material is placed in a 50 cm³ stoppered sample tube to which are added 4 cm³ acetone. After vigorous shaking for 10 s and allowing to stand for 10 min, 3 cm³ water are added and the mixture is shaken once more. Finally, 3 cm³ petroleum ether are

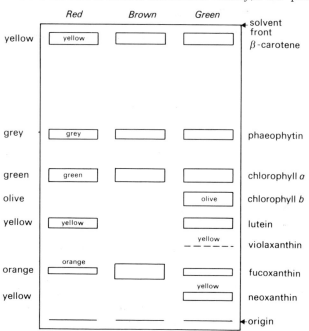

Figure B8.10 Typical appearance of a thin-layer chromatographic plate obtained with pigment extracts from brown, green and red algae.

added and after 5 s of vigorous shaking the solvents are allowed to separate. The pigments will practically all collect in the top layer of petroleum ether and this can now be withdrawn with a pipette for further examination by either thin-layer chromatography or spectrophotometric analysis.

(B) ABSORBANCE

The separated pigments should be examined under natural (visible) and 340 nm (3400 Å) ultraviolet light. Individual pigment bands can be scraped off with blades and the coloured silica gel collected in 10-cm^3 centrifuge tubes containing 4 cm^3 of 80 cm^3 acetone diluted with 20 cm^3 water. After mixing the contents of the tubes, they are centrifuged for 2 min at 750 **g** and the supernatant is decanted for absorption spectra analysis in a manual or recording spectrophotometer between 350 and 720 nm (3500 and 7200 Å) for chlorophylls and between 350 and 600 nm (3500 and 6000 Å) for carotenoids.

(C) TRANSMISSION AND REFLECTION

To demonstrate that the appearance of colour as sensed by our eyes depends on the wavelengths of light transmitted and reflected by the substance under observation, it is useful to shine beams of white, green, red and blue light through a solution of chlorophyll in acetone. As red and blue light are absorbed, the solution will appear to be dark.

(D) TRANSMISSION SPECTRA

A quantaspectrometer is analogous to a spectroradiometer measuring the quantum flux at intervals between 400 and 700 nm (4000 and 7000 Å) across the spectrum. By placing the plant material across the sensor, the instrument will measure its transmission spectrum.

If available, a quantaspectrometer should be used to obtain transmission characteristics of both the extracted pigments and the fresh material. In respect of the algal extracts, measurements can be related to hypotheses of 'chromatic adaptation' to wavelengths available at different depths.

(E) FLUORESCENCE

The most primitive method of showing fluorescence of extracts of photosynthetic pigments employs filtered extracts of fresh grass-cuttings steeped for a few minutes in acetone. The filtrate should be of a rich, clear, green colour when contained in a test-tube or a 250 cm^3 measuring cylinder. When 50-cm^3 conical flasks are completely filled with such an extract and stoppered, the solution will appear blood-red or brown if illuminated by a bench lamp or near a window. Gentle swirling and shaking and inclining the flask away from the vertical will enhance the effect; the solution will momentarily appear green.

(F) FLUORESCENCE IN ULTRAVIOLET LIGHT
(Based on information given by Dr K. Hardwick, University of Liverpool.)

The fluorescence at 340 nm (3400 Å) ultraviolet light of the extracts obtained as for Experiment 8.10 and that of the original fresh plant material should be measured and compared.

Illuminated fresh material viewed through Wratten gel filters No. 97 (catalogue no. 187 7083) (from Kodak, see Part C2) shows red fluorescence very well.

CAUTION: Safety spectacles should be used when experimenting with ultraviolet light.

9 Translocation

9.1 Xylem translocation
By Professor H. Meidner, University of Stirling
cf. Expt 5.1

Several aspects of translocation of water and solutes in the xylem are dealt with in Experiments 4.1, 4.4 and 5.1–5.3. Standard, well known and well tried experiments involving xylem translocation can be found in many texts (Meidner & Sheriff 1976; most standard texts) and therefore require no recapitulation here. Only a few techniques will be presented which can form the basis for xylem investigations.

(A) IDENTIFICATION OF XYLEM TISSUE

In anatomical studies stains specific to xylem tissue (saffranin, see Part C1) can readily be used on transverse and longitudinal sections, and in preparations such as described in Experiment 5.1; in macerates (see Part A) xylem vessels can be recognised by their wall pattern without the use of stains. However, for the identification of the path of translocated liquid, non-specific dyes (e.g. eosin or ink) are useful with translucent stems alone, e.g. busy Lizzy (*Impatiens sultani*) or in combination with serial, transverse hand-cut sections of *Coleus blumei*. The xylem paths at nodes are often less known and worth exploring. The major disadvantage of most dyes is that they are adsorbed and thus do not travel at the same rate as the liquid in which they are dissolved. This makes them unsuitable for studies of rates of xylem translocation but does not interfere with experiments identifying the xylem tissue.

(B) TRANSPIRATION PULL

Prepared woody twigs with leaves (see Part A) placed in an assembly as shown in Figure A3b are suitable material for demonstrating the magnitude of the transpiration pull or the existence of negative hydrostatic pressures in the xylem of transpiring plant parts.

(C) RATE OF XYLEM TRANSLOCATION: QUALITATIVE DEMONSTRATION

A qualitative demonstration of the rate of xylem translocation can be made with lupin leaves (*Lupinus alba*, or cultivated spp.). A 15- to 20-cm-long petiole should be cut under water (see Part A) and, while illuminated in a gentle air current, the leaves are allowed to wilt visibly

Suitable species of plants must be found by trial and error in each locality. The widely distributed cherry laurel (*Prunus laurecerasus*) can be recommended.

The twig should be illuminated and exposed to gentle air movements. The mercury should be added once stomata are open and transpiration is proceeding (see Fig. A3b). A properly prepared twig will readily raise the mercury column to a height of 0.38 m, representing a 5.00-m-high column of water or a negative hydrostatic pressure potential of − 0.05 MPa.

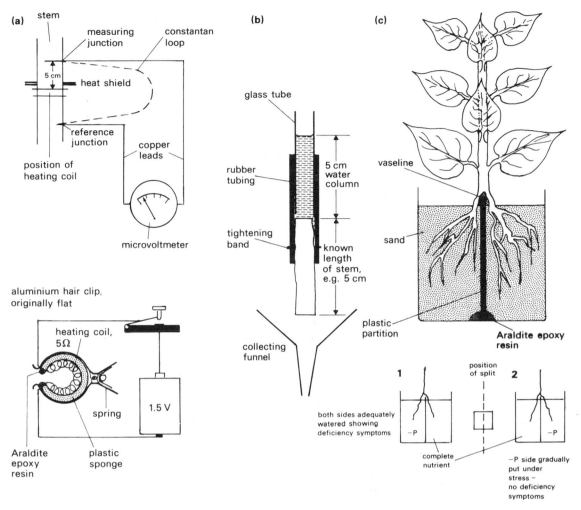

Figure B9.1 (a) Transverse section of clip used for heating plant stems and circuit diagram for its operation. (b) Arrangement suitable for measuring the hydraulic conductivity of stem tissue. (c) Longitudinal section of vessel for split root system using *Coleus blumei*. The plane of splitting the root system is shown in plan and treatments 1 and 2 are indicated schematically.

by withholding water. A water supply is then restored to them by cutting off 5 cm of the base of the petiole under water. Turgor will be regained speedily and the full display of these compound leaves is visibly reached within 2 min. By measuring the length of the petiole (about 15 cm) and the area occupied by xylem tissue in a transverse section of the petiole, together with the increase in weight of the leaf as it recovers turgor, the rates of liquid flow per square centimetre can be calculated.

If leaf water potential measurements can be made using a pressure bomb (Expts 5.7 & 5.8) or psychrometer (see Expt 5.4) when the leaf is wilted and when fully turgid (at which point leaf water potential should be zero), the force driving the water into the wilted leaf can be calculated.

(D) RATE OF XYLEM TRANSLOCATION: QUANTITATIVE MEASUREMENT

A quantitative measurement of the rate of xylem translocation can be made with the heat pulse method using the home-made assembly shown in Figure B9.1a. Round stems are most suitable: *Xanthium strumarium*, *Ricinus communis* and *Pelargonium zonale* are recommended.

(E) CONDUCTING CAPACITY

Stems of 5 cm length, cut off under water, are prepared from *X. strumarium* or *R. communis* and from climbing plants such as *Hedera helix*. These stem sections are assembled in the apparatus shown in Figure B9.1b. Measurement of the area occupied by xylem tissue and of the diameters of xylem vessels in transverse sections of the stems allows for the conductance to be calculated and expressed in $cm^3 cm^{-2} min^{-1}$. If herbaceous stems are used, e.g. cucurbitaceous climbers, some matrix will be needed to effect a seal. In some woody species it will be necessary to remove the phloem (bark) to avoid 'gumming' of the wood.

(F) WATER STRESS AND XYLEM SOLUTE TRANSLOCATION

It usually takes 4–6 weeks for cut shoots of *Coleus blumei* to develop good root systems in compost and 3–4 weeks for the phosphorus-deficient treatment effects to become noticeable.

If desired, cut shoots can be rooted in nutrient solution and part of the experiment can be carried out with the split root system in complete and phosphorus-deficient nutrient solutions, but the third treatment with water stress is more difficult to apply than with soil-rooted plants.

Coleus blumei or *C. frederici* plants with two pairs of opposite decussate leaves, grown in a medium that allows their removal without undue damage to the root system, are used for this demonstration. The root system must be gently washed to remove most of the original growing compost before the basal 2 cm of the stem and the whole root system are split with a sharp blade down the middle of the square stem (not diagonally from corner to corner). The two parts of the root system are planted in acid-washed sand contained in a subdivided container as shown in Figure B9.1c. The inside surfaces of the split stem straddling the edge of the dividing septum should be coated with Vaseline to prevent drying out.

There are three treatments, each given in duplicate. Six plants are required. One treatment consists of supplying both half root systems of two plants with complete nutrient solution (see Part C1). The second treatment uses two plants with the complete nutrient solution on one side and a phosphorus-deficient one (−P) on the other. The third treatment begins in the same way as the second, but gradually withholds watering with the phosphorus-deficient solution from the two plants.

The first treatment will allow normal growth with typically coloured, symmetrical leaves. The second treatment will result in a plant with smaller and probably paler leaves above the phosphorus-deficient medium; some leaves may show the symptoms asymmetrically on the two sides of the mid-vein. The third treatment where the phosphorus-deficient feeding has been withheld will not show asymmetry, although a less vigorous growth all round may be seen. The leaves on the phosphorus-deficient side may wilt slightly when the watering is gradually withheld, but will recover as water and phosphorus are drawn across the stem by the low water potential developing on this side.

REFERENCES

Baker, D. A. and J. A. Milburn 1965. *Nature 205*, 306–7.
Meidner, H. and D. W. Sheriff 1976. *Water and plants*. London: Blackie.

RELEVANT ANATOMICAL STUDIES

When determining the hydraulic conductance of different stems (e.g. *Hedera helix*, *Cissus capense*, *Cucurbita pepo*, *Cucumis sativus*, *Xanthium strumarium* and *Ricinus communis*), the xylem vessel diameters can

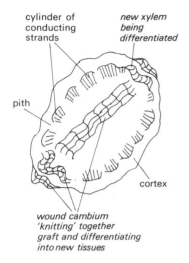

cylinder of
conducting
strands

*new xylem
being
differentiated*

pith

cortex

*wound cambium
'knitting' together
graft and differentiating
into new tissues*

Figure B9.1d Plan of a transverse section through the graft union of two tomato plants.

be measured with eyepiece graticules (see Part A) using thin, hand-cut sections with or without saffranin staining. In the cucurbitaceous stems, bicollateral phloem may also be observed.

TRANSLOCATION IN GRAFTED STEMS

A most instructive use of dyes and sections is in combination with an exercise in approach grafting. Two young tomato plants of about 12–15 cm height and with stems of about 0.4 cm diameter are used. Both plants must grow in the same pot; the stems are cut longitudinally at a height of about 6–8 cm above soil level, so that for the length of 1–1.5 cm one-third of the tissue facing each other are removed. The two cut surfaces are then placed in contact with each other and held together to be taped with *dry* Sellotape. A support for the two plants should be provided from the time of the grafting operation. Within 2 weeks the graft should have taken and one of the stems can be cut below the graft. Two days later the shoot of the stock can be cut off, allowing the scion to obtain its sap from the root stock.

When the grafted plant has grown for a further 2 weeks, transverse sections of the graft region will show the regeneration of the different tissues by wound cambium formation (cf. Expt 2.4 anatomical studies) on either side of the 'seam' of the cut as indicated in Figure B9.1d.

If the stock of a successful graft union is cut off above soil level and placed in a solution of eosin dye while the scion transpires, the path of water flow in the xylem will become stained. Longitudinally cut stems will show the 'cross-over' paths of xylem within the graft union and longitudinal sections can be prepared for more detailed studies.

9.2 Phloem translocation
**Based on information given by Dr H. A. Collin, University of Liverpool
cf. Expts 4.3, 4.4, 4.6, 4.8, 4.9, 9.5 & 11.3**

(A)   RATE OF PHLOEM EXUDATION

The occurrence of phloem turgor pressure can be recognised by phloem sap exudation from cut petiole stumps; the recommended species is squash (*Cucurbita pepo*). Plants should be 10–12 weeks old and strong, about 1.5 m high with 2- to 3-cm-diameter stems. At the time of experimenting, the plants must not be under conditions promoting the development of root pressure (see Expt 5.1) as this would result in the xylem sap diluting the phloem exudate and alter the rate and composition of the measured exudate.

Immediately after cutting off an upper leaf 1–2 cm up from the base of the petiole, the exudate is collected from the petiole stump attached to the stem with a 10 mm³ glass capillary. By measuring the length of the column in the capillary, its volume can be estimated before the exudate is transferred to a small sample tube for analysis (see below). As the sap gels in air, a fresh cut 0.1 cm further down the petiole stump should be made every 30 s and the exudate collected. The average rate of exudation over 3–5 min can then be calculated.

Comparisons can be made between leaves from plants illuminated with very weak and strong light sources and between young, mature and senescent leaves, but in all cases leaf areas should be comparable.

Rate measurements should be expressed as cm^3 (sap exuded) cm^{-2} (phloem cross-sectional area) s^{-1}. Phloem area estimates can be arrived at using calibrated eyepiece graticules (see Part A) on hand-cut sections of the petiole stump.

(B) COMPOSITION OF PHLOEM SAP

Qualitative tests

Sucrose
Sucrose identification is carried out by hydrolysis of exudate with 1 drop of 2×10^3 mol m^{-3} (2.0 M) HCl at 70 °C for 30 min. 'Diostix' with colour chart for glucose or 'Clinitest' tablets (from Miles Laboratories, see Part C2) will give an indication of glucose content and total reducing sugar levels, respectively.

Amino acids
Five cubic millimetres of exudate are spotted on to a filter paper and sprayed with a ninhydrin preparation (see below). After 10 min the presence of amino acids will be indicated by a violet-blue colour.

Potassium
The presence of potassium is shown by placing 1 drop of exudate on a 'Merckoquant strip' (from BDH Chemicals Ltd, see Part C2) specific for K^+ and comparing this with the colour chart.

Ninhydrin

The ninhydrin preparation is made from two solutions, A and B:

solution A is 0.16 g $SnCl_2$ in 500 cm^3 citrate buffer at pH 5.0 (see Part C1);
solution B is 4.0 g ninhydrin in 100 cm^3 2-methoxyethanol.

Equal portions of A and B are mixed immediately before use.

Quantitative estimations

Sucrose
Between 20 and 30 mm^3 exudate are first dissolved in 1 drop of 60 cm^3 ethanol diluted with 40 cm^3 water. The drop is then made up to 1 cm^3 with diluted ethanol. This is added to 2 cm^3 anthrone (see below) in such a manner that the dissolved exudate floats above the anthrone, which has previously been cooled to 10 °C. This mixture is then chilled on an ice–water mixture to 10 °C after which the two layers are gently mixed and heated for 16 min at 90 °C in a water-bath. On cooling to room temperature, the presence of glucose is indicated by a blue-green colour. The absorbance at 625 nm (6250 Å) of the coloured solution can now be measured in a calibrated spectrophotometer or a colorimeter using a red filter.

For the calibration of the spectrophotometer, a range of sucrose solutions at 0.0, 0.2, 0.4, 0.6, 0.8 and 1.0 mol m^{-3} (0.0–1.0 mM, in steps of

0.2 mM) in diluted ethanol (60 cm³ ethanol and 40 cm³ water) will be needed, and 1.0 cm³ of each solution is added to 2.0 cm³ anthrone treated as above.

Amino acids

Between 30 and 40 mm³ exudate are dissolved in 0.5 cm³ dilute *n*-propanol (50 cm³ *n*-propanol plus 50 cm³ water). Of this diluted exudate 0.1 cm³ is then added to 1.0 cm³ ninhydrin preparation (see Qualitative tests) and, after mixing, heated for 10 min in a boiling water-bath. Next, the preparation is diluted with 5 cm³ of the diluted *n*-propanol, shaken, and left for 15 min before absorbance is measured at 500 nm (5000 Å) in a calibrated spectrophotometer or in a colorimeter using a green filter.

For the calibration a solution of glutamic acid (see below) can be used. Dilutions to 0.0, 10, 12, 40, 60, 80 and 100 μg 0.1 cm⁻³ in diluted *n*-propanol (50 cm³ propanol in 50 cm³ water) should be prepared.

Potassium

Twenty cubic millimetres of exudate are added to 1 cm³ dilute ethanol (60 cm³ ethanol plus 40 cm³ water); if a white precipitate results, this must be allowed to settle out, or the mixture must be centrifuged. The supernatant is used in a flame photometer which should be calibrated using 1 mol m⁻³ KCl in steps of 0.2 mol m⁻³.

ANTHRONE

The anthrone is prepared as follows: 760 cm³ concentrated H_2SO_4 are slowly added to 300 cm³ distilled water (CAUTION). After cooling to room temperature, 1.5 g anthrone are dissolved in the diluted acid.

GLUTAMIC ACID

The solution for calibration is prepared by dissolving 100 mg glutamic acid in 100 cm³ dilute *n*-propanol (50 cm³ *n*-propanol plus 50 cm³ water).

RELEVANT ANATOMICAL STUDIES

The cortex or 'bark' of stems can be studied in thin, hand-cut sections, but better in microtome sections. Castor bean (*Ricinus communis*), which is used in Experiments 4.1, 4.4, 5.3 and 9.3, provides suitable material for the demonstration of phloem tissue with phloem parenchyma and fibres, cortex parenchyma, *cork cambium* and cork with lenticels (cf. Expts 10.2, 10.8; & 2.4 anatomy; abscission is accompanied by peridermal wound cambium activity).

REFERENCE

Colowick, S. and N. O. Kaplan (eds) 1957. *Methods in enzymology*, vol. III. New York: Academic Press.

9.3   Carbohydrate distribution by phloem translocation
Based on information given by Professor D. A. Baker, Wye College

Two plants of the same species, either of squash (*Cucurbita pepo*) or castor bean (*Ricinus communis*), are prepared for $^{14}CO_2$-feeding for 1 h by arranging a planchette held by a wire holder (Fig. B9.3) near the lamina of one leaf about half way up the stem. This leaf is then enclosed in a plastic bag, which must be sealed around the petiole and the stem of the wire holder with foam plastic compressed by metal ties. A droplet containing 20 μCi NaH$^{14}CO_3$ solution (see Expt 8.7) is injected on to the planchette followed by a larger drop of 10^3 mol m^{-3} (1.0 M) HCl; thereafter, the syringe needle hole must be sealed with Sellotape.

After 1 h the bag is removed in a fume cupboard and one plant placed in darkness while the other is illuminated for 3 h. Plants should be labelled D and L.

After 3 h, samples of exudate are collected from the middle of internodes and petioles (Fig. B9.3) after cutting with a new blade and drying immediately with tissue paper. With *R. communis* the internode is not cut through, but to obtain exudate the 'bark' is incised to a depth of about 0.1 cm. When a drop of exudate has formed, it is touched into a weighed and labelled planchette, dried in an oven and the $^{14}CO_2$ activity counted for 2–5 min. Results are expressed as counts min^{-1} mg^{-1} dry wt exudate and class results should be pooled.

9.4   Carbohydrate analysis of phloem exudate
Based on information given by Professor D. A. Baker, Wye College
cf. Expt 11.5

For this exercise three squash (*Cucurbita pepo*) and three castor bean (*Ricinus communis*) plants must be prepared by $^{14}CO_2$ assimilation for 30 min, 2 h prior to use in practical classes.

^{14}C-labelled exudate from cut stem internodes is collected in a microcapillary tube. By carefully inserting a 5 mm^3 (5 μl) micropipette into the capillary tube, two 5 mm^3 volumes are withdrawn and deposited 2 cm apart on the starting line of a thin-layer chromatographic plate. Deposition should be carried out by repeatedly touching the Kieselguhr with the micropipette so that the areas of the two spots remain as small as possible. Between the repeated applications, each deposit should be allowed to dry in a warm air current from a hair drier.

Leaving a 4 cm gap between these two spots and the next series of six spots, 5-mm^3 samples are taken from six solutions of glucose, fructose, galactose, sucrose, raffinose and stachyose (prepared in each case by dissolving 1 g in 100 cm^3 water) and these are placed on the starting line in the same manner as the two exudate spots.

The prepared plate is carefully transferred to a tank containing formic acid, butanol, tertiary ethyl alcohol and water in the proportions 15:30:40:15 by volume, the spots must remain above the solvent surface. When the solvents have run 10 cm the plate is dried in a fume cupboard. If necessary, two runs may be made to increase the R_f values (the R_f value is the ratio of the distance travelled by the compound to the distance travelled by the solvent front). Once dry, the plate is sprayed

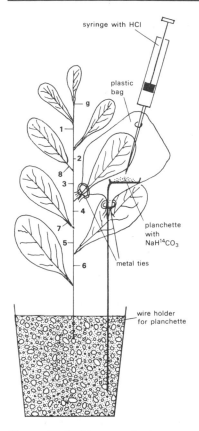

Figure B9.3 Diagram of plant prepared for $^{14}CO_2$-feeding of one leaf. The method of sampling exudate from squash (*Cucurbita pepo*) is as follows: the middle of internodes 1–6 and petioles g, 7 and 8 must be cut with clean razor blades and quickly blotted with tissue paper. When thereafter exudate appears on the cut surface of the plant (not the cut-off portion), it is touched on to a prepared and weighed planchette. This must be dried in an oven before counting for $^{14}CO_2$ for 2–5 min. Results are expressed as counts min^{-1} mg^{-1} dry wt.

Experiments involving the use of radioactive material must be carried out according to the safety regulations in force. All radioactive waste, including plant material, must be disposed of immediately after the conclusion of experiments.

with p-anisidine–HCl (see below) and then dried for 5 min in an oven at 100 °C, when the different sugars will appear in their characteristic positions and colours.

^{14}C-COUNTING

The separated sugars in the two columns from the two samples of exudate can be scraped off individually and combined in one planchette for each sugar. Counts per minute per sample can be made when the planchette content is dry. A 5 mm^3 sample of the original exudate should be dried on a planchette for counting and compared with the summed counts from the sugars used in the chromatogram.

p-ANISIDINE–HCl

To prepare p-anisidine–HCl, 2 g p-anisidine are dissolved in 5 cm^3 concentrated HCl; to this are added 10 cm^3 concentrated orthophosphoric acid and the mixture is made up to 200 cm^3 with ethanol.

ASSOCIATED ANATOMICAL STUDIES

Using unlabelled plants, 5- to 8-cm-long sections of squash shoots are obtained for observation of the apical end of the section under low magnification. The positions and the identities of the different tissues, especially the phloem (bicollateral) should be mapped out (cf. Expt 9.1 anatomy).

A drop of 1% mercapto-ethanol in 50 cm^3 acetate buffer (see Part C1) at pH 5.6 is placed on a *coverslip*. After cutting through the stem in the middle of the apical internode of the 5- to 8-cm-long shoot section, the cut surface should be dried with paper tissue. A droplet of exudate will be seen to collect gradually. The exudate should be touched very gently into the drop of buffer solution on the coverslip and slowly raised to a microscope slide held above it until the droplet touches the slide and the coverslip adheres to it. *No pressure whatever must be exerted*. The microscope slide with the coverslip adhering to it is then quickly turned over so that the coverslip is uppermost. By examination under 400× magnification, intracellular strands and other phloem inclusions can be seen. The strands will stain brown with straw-coloured iodine solution (see Part C1) and allow for measurements of dimensions using calibrated eyepiece graticules (see Part A).

9.5 Uptake and translocation of radioactively labelled sulphate ions in plants
Based on information given by Dr J. Hannay, Imperial College of Science and Technology
cf. Expts 4.3, 4.4, 9.2 & 11.3

Ten *Phaseolus vulgaris* plants are required with primary and first trifoliate leaves. Five plants are placed for 24 h in each of two 200 cm^3 vessels containing nutrient solution (see Part C1). Then they are given a 1 h treatment with $^{35}SO_4^{2-}$ by the addition to the nutrient medium of 0.2 cm^3

of a solution of $Na_2^{35}SO_4$ containing 200 μCi cm^{-3} (40 μCi). During this hour the plants must be kept in a well ventilated and sunny place.

After the radioactive treatment plants are removed from the vessels, their roots thoroughly washed and immediately placed in vessels containing 200 cm^3 fresh non-radioactive culture solution. Two plants each are prepared for autoradiography at times 0, 6, 24, 48 and 96 h after completion of the 1 h feeding time.

Two methods of preparing plants for autoradiography can be used: either 'freeze drying', which prevents subsequent isotope movement; or blotting between paper, when it is necessary to cut off parts from the stem with a sharp blade and label them *before* drying for 3 h in a preheated photographic print drier. Plant parts must not lie on top of each other. After reassembling the parts, they are placed on an X-ray film in an autoradiographic folder for 14 days and then developed.

The autoradiographs will show rates of isotope movement and its distribution during the 4 days; if possible, an extension to 6 days or an initial feeding of up to 12 h could be tried to give more definite results.

By introducing 'ringing' treatments, the identity of the xylem intake and export paths, as distinct from the phloem path, can be demonstrated. For intake, the ringing must be carried out before the isotope feeding; for export, the ringing must follow the feeding. Three methods of ringing can be used:

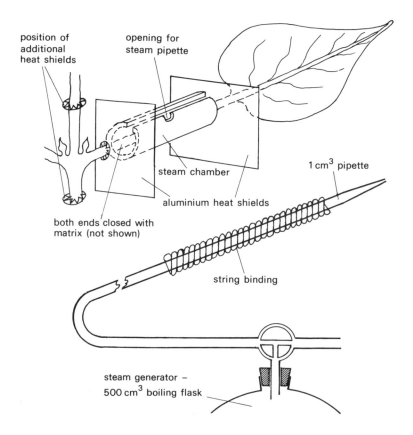

Figure B9.5 Longitudinal section of steam generator and steam chamber for 'girdling' petioles. The rest of the plant must be protected by aluminium foil sheets on either side of the steam chamber and above and below the insertion of the petiole on the main stem. The treated leaf must be supported by foam plastic in its horizontal habit.

(a) Steam ringing of petioles by a system as shown in Figure B9.5. The steam is applied to the chamber for 1 min in the case of *P. vulgaris*; for other species different times may be required. The heat shields and the support for the leaf are essential. Temporary wilting of the treated leaf may occur and the technique should be practised on non-radioactive plants; treated leaves should regain and maintain turgor for 48 h.

(b) Surgical removal of the cortex.

(c) Dry burning by the device described in Experiment 9.1d.

Methods (b) and (c) are not suitable for herbaceous stems or petioles, but only for woody stems.

10 Hormones

10.1 Extraction and purification of indole-3-acetic acid (IAA), abscisic acid (ABA), gibberellic acid (GA$_3$) and cytokinins

Based on information given by Professor M. A. Hall and Dr A. R. Smith, University College, Aberystwyth
cf. Expts 2.1, 2.2, 10.2, 10.3, 10.5 & 10.10

Extracts of ABA and IAA from over-ripe avocado fruit (*Persea gratissima*, ABA) and cabbage heart (*Brassica oleracea* var. capitata, IAA) can be obtained by grinding 50 g tissue in a chilled mortar, and then transferring to 100 cm³ diethyl ether in a conical flask standing in ice for 1 h, occasionally swirling the liquid round.

Twenty-five cubic centimetres of a solution of 5 g sodium bicarbonate in 100 cm³ water are then added to the mixture placed in a separating funnel. After gently shaking, the lower aqueous layer is run off and retained. The upper ether layer is once more washed with 25 cm³ of the NaHCO$_3$ solution and the two aqueous layers so obtained are combined. The pH is adjusted to 3.0 with 2.5×10^3 mol m^{-3} (2.5 M) H$_2$SO$_4$ and the mixture is placed in a separating funnel for extraction with three successive 50 cm³ volumes of diethyl ether. The combined ether extracts, about 150 cm³, are dried with anhydrous MgSO$_4$, filtered off and evaporated to dryness, and the residue is dissolved in 2 cm³ diethyl ether.

This procedure is also suitable for the extraction of GA$_3$ (cf. Expt 10.5) from commercial frozen peas (*Pisum sativum*) and of cytokinins from coconut milk (*Cocos nucifera*).

10.2 Separation of extracts of indole-3-acetic acid (IAA), abscisic acid (ABA), gibberellic acid (GA$_3$) and cytokinins

Based on information given by Professor M. A. Hall and Dr A. R. Smith, University College, Aberystwyth
cf. Expts 10.1, 10.4 & 10.10

The 28% NH$_3$ used in the solvent mixture is prepared from a solution of 1 part ammonia solution of sp. gr. 0.88 in 3 parts water. The final concentration of ammonia in the solvent mixture is 3×10^2 mol m^{-3} (0.3 M).

Each extract obtained in Experiment 10.1 is loaded on to Whatman's 3 MM chromatography paper as a 1-cm-wide, 10-cm-long baseline, 2–3 cm up from the lower edge. After allowing the extract to dry, the sheet is placed as a descending chromatogram in a tank overnight with a solvent mixture of iso-propanol–28% NH$_3$–H$_2$O (10 : 1 : 1 by volume).

It is important to mark each strip with its R_f value.

After marking the solvent front, the chromatogram is dried thoroughly in a fume cupboard and cut into 10 equal transverse strips between R_f 0.0 and 1.0, with an additional strip from below R_f 0.0 (the R_f value is the ratio of the distance travelled by the compound to the distance travelled by the solvent front). Also cut several strips that have been impregnated with hormone solutions of known concentrations:

IAA, between 2×10^{-2} and 2×10^{-4} mol m^{-3} (2×10^{-5} M and 2×10^{-7} M)
ABA, 10^{-2} to 10^{-4} mol m^{-3} (10^{-5} to 10^{-7} M)
GA$_3$, 10^{-1} to 10^{-5} mol m^{-3} (10^{-4} to 10^{-8} M)
kinetin, 10^{-2} to 10^{-6} mol m^{-3} (10^{-5} to 10^{-9} M)

10.3   Wheat coleoptile straight growth test
Based on information given by Professor M. A. Hall and Dr A. R. Smith, University College, Aberystwyth
cf. Expts 10.1, 10.2 & 10.4

After soaking wheat grains with husks (caryopses of *Triticum vulgare*) in water at room temperature for 2 h they are placed in moist vermiculite at 25 °C in darkness. On about day 3, 2-cm-long coleoptiles can be cut into 1-cm-long sections under a green safe light by using a double edged, fixed distance cutter. The last 0.3 cm of the coleoptile tip should not be included, thereby removing the source of endogenous hormone. For cutting the coleoptile sections, the use of a glass slide with two lines ruled 2 cm apart is recommended. The coleoptiles are then placed for cutting with the double edged cutter so that 0.3 cm tips extend beyond one of the lines. The sections are kept in distilled water prior to use in the assay.

Each of the 10 chromatography paper strips obtained from the separations (Expt 10.2), as well as those impregnated with known indole-3-acetic acid (IAA) solutions, are rolled so that they fit against the inside walls of 1.7-cm-diameter stoppered glass vials.

To each vial are added 1.2 cm^3 of a buffered sucrose solution (see below), and the vials are rotated in a vertical orbit for 2 h so that the hormone will elute. Next, a random sample of the 1-cm-long coleoptile sections is placed into the vials adhering to the paper and the assembly is rotated for another 20 h. If a rotating rack is not available, the test can be carried out in Petri dishes, but this may result in curved growth. At the conclusion of the experiment, coleoptile lengths are measured and presented in the form of histograms; for significant differences an analysis of variance will be required.

Wheat coleoptiles will elongate in buffered sucrose solutions without any added IAA; it is therefore possible to modify this assay to detect compounds in extracts which can overcome the inhibitory effect of abscisic acid (ABA). For this modification the buffer is prepared with a solution of ABA instead of distilled water, resulting in a final concentration of 0.1 g m^{-3} ABA.

For the ABA experiment the 10 times dilution is with ABA, not distilled water. The ammonium salt is given by dissolving 1 mg ABA in a minimum of ammonium hydroxide; this is made up to 90 cm^3 with distilled water, and diluted 1 : 100 to give a concentration of 0.001 mg ABA in 9 cm^3 of solution. The concentrated buffer prepared for the wheat assay must now be diluted with ABA solution to give an ABA concentration of 0.1 g m^{-3}. This is accomplished by diluting 10 cm^3 of the concentrated buffer with 90 cm^3 of the dilute ABA solution prepared as above.

BUFFERED SUCROSE SOLUTION

This is prepared using 4.485 g K$_2$HPO$_4$ plus 2.547 g citric acid monohydrate in 250 cm^3 distilled water. The solution is diluted 10 times,

5 g sucrose are added and the pH is adjusted to 5.3 with 10^3 mol m^{-3} (1.0 M) HCl.

RELEVANT ANATOMICAL STUDIES

The anatomy of coleoptiles of cereal seedlings should be investigated while working with this material. Dissection is accomplished with a pair of forceps and a needle to reveal the hollow structure of the coleoptile and the point of attachment of the first leaf to the developing shoot axis. A few transverse sections of the structure will show the folding and rolling of the leaves.

10.4   Oat coleoptile bioassay
Based on information given by Professor M. A. Hall and Dr A. R. Smith, University College, Aberystwyth
cf. Expts 10.2 & 10.3

This bioassay gives less variation in individual coleoptiles than when wheat is used, and generally straight growth.

All initial preparations are as for Experiments 10.2 and 10.3, but in place of wheat, oat grains (*Avena sativa*) are used; these need 84 h growth in vermiculite under continuous dim red light at 25 °C. Coleoptile sections are prepared under the green safe light as for wheat.

The R_f zones of the chromatogram (R_f value being the ratio of the distance travelled by the compound to the distance travelled by the solvent front) are cut into 5-cm-diameter Petri dishes and eluted with 4 cm^3 distilled water for 3 h. Ten coleoptile sections are then placed in the dishes and incubated at 25 °C for 1 h with a gentle rocking motion. After this, 1 cm^3 of a solution of 0.45 g glucose plus 0.17 g KH$_2$PO$_4$ in 100 cm^3 water at pH 4.8, adjusted with 10^3 mol m^{-3} (1.0 M) HCl or NaOH, is added and the sections are incubated for another 5 h before measurements are carried out.

For purposes of calibration, 1 cm^3 of a solution of 0.87 mg indole-3-acetic acid (IAA) in 100 cm^3 distilled water is added to the control bioassay Petri dishes, resulting in a final concentration of 10^{-2} mol m^{-3} (10^{-2} μM) abscisic acid (ABA). (cf. Expt 10.3 anatomy.)

10.5   Lettuce hypocotyl extension bioassay for gibberellic acid (GA$_3$)
Based on information given by Professor M. A. Hall and Dr A. R. Smith, University College, Aberystwyth
cf. Expts 10.1, 10.2 & 10.10

Good quality lettuce seeds (*Lactuca sativa* L.) should be placed on moist paper (Whatman 3 MM) in plastic boxes of roughly 30 × 16 × 10 cm; the boxes are kept in darkness at 20 °C. After 36–40 h seedlings of similar size are selected for the experiment and ten of these used in each treatment with the elutions from the different R_f zones plus 3 cm^3 one-quarter strength nutrient medium.

After preparing the extracts and R_f fractions from frozen peas as described for Experiments 10.1 and 10.2, the 10 chromatogram strips are placed in 4-cm-diameter Petri dishes and 10 lettuce seedlings are put into each dish with 3 cm^3 of one-quarter strength nutrient medium (see Part C1). The material is incubated at 25 °C for 3 days in continuous light. Measurements of hypocotyl length are made with callipers; if it is necessary for reasons of time to delay measurement, dishes can be kept for several days in a cold room at 4 °C, when hypocotyl growth is minimal and does not markedly affect the bioassay.

For calibration a series of GA$_3$ solutions are prepared of between 10^{-1} and 10^{-5} mol m^{-3} (10^{-4} to 10^{-8} M) GA$_3$ and, of each strength, 3 cm^3 are

The lettuce hypocotyl assay is sensitive to about 4×10^{-3} µg GA_3 but the effects are readily inhibited by compounds such as ABA present in impure extracts and masking GA_3 activity. It is important therefore to take care to avoid cross contamination while preparing serial dilutions and when transferring seedlings. It is best to work from dilute to more concentrated solutions.

used per Petri dish. Each dish is lined with filter paper and contains 10 lettuce seedlings. The serial dilutions are best prepared by making a 10^{-1} mol m^{-3} (10^{-4} M) GA_3 stock solution with one-quarter strength nutrient medium (see Part C1); 1 cm^3 of this stock solution is then added to 9 cm^3 nutrient medium and properly mixed giving the 10^{-2} mol m^{-3} (10^{-5} M) GA_3 solution. This method is repeated to give the 10^{-3}, 10^{-4} and 10^{-5} mol m^{-3} (10^{-6}, 10^{-7} and 10^{-8} M) strengths, always using the nutrient medium for the dilution. For the calibration, mean lengths of the 10 hypocotyls in each treatment are plotted against $\log_{10}$ GA_3 concentration.

All treatments should be carried out in duplicate, including those without GA_3 and the calibration treatments. In addition to the calibration curve, histograms should be prepared for each R_f zone elution.

10.6 *Amaranthus* cytokinin bioassay

Based on information given by Professor M. A. Hall and Dr A. R. Smith, University College, Aberystwyth

Cytokinins stimulate cell division, cell enlargement and seed germination; they counteract apical dominance and retard leaf senescence. This bioassay, however, is based on the cytokinin-induced formation of betacyanin in cotyledons and hypocotyls of *Amaranthus caudatus* incubated in the presence of tyrosine in the dark.

After the removal of the seed coats, seeds are germinated in plastic boxes as in Experiment 10.5 for 72 h at 25 °C. Explants of upper portions of hypocotyls with the two cotyledons are prepared using new, single edged blades. Ten explants with the known cytokinin standards (see below) are placed in each of the bioassay boxes, together with those without cytokinin. All treatments must be replicated 3 times. The explants are incubated for 18 h at 25 °C in the dark and then each batch of ten is transferred to a test-tube containing 3 cm^3 distilled water, frozen, thawed, refrozen and thawed again to disrupt the cells (cf. Expt 3.4). After removing the explants, the 3 cm^3 betacyanin solution from each treatment are measured for their absorbance in a spectrophotometer at 542 and 620 nm (5420 and 6200 Å) (cf. Expt 3.4). Since over a wide range the concentration of betacyanin is proportional to the difference between absorbance at 542 and 620 nm (5420 and 6200 Å), this difference is plotted against $\log_{10}$ molar concentration of the standard cytokinin solutions. The results obtained with the elutions from the different R_f zones can then be interpreted (the R_f value is the ratio of the distance travelled by the compound to the distance travelled by the solvent front).

The bioassay boxes are transparent polystyrene boxes, $5.6 \times 3.5 \times 2.2$ cm. They are lined with two layers of Whatman 3 MM paper moistened with 2 cm^3 of 7.5×10^2 mol m^{-3} (0.75 M) phosphate buffer (cf. Expt 10.3) containing 1 cm^3 of a solution of 1 g tyrosine in 1000 cm^3 water and, in the case of the experimental material, the elutions from the R_f zone chromatogram.

CYTOKININ STANDARDS

Suitable cytokinin standard dilutions can be prepared as for gibberellic acid (GA_3) (Expt 10.5), starting with 10^{-2} mol m^{-3} (10^{-5} M) and diluting 4 times to obtain 10^{-3}, 10^{-5} and 10^{-6} mol m^{-3} (10^{-6}, 10^{-8} and 10^{-9} M) cytokinin.

10.7   Abscission and ethylene
Based on information given by Dr R. Sexton, University of Stirling
cf. Expts 2.4 anatomy, 2.5, 10.8 & 10.9

Other suitable material for these experiments are explants of *Coleus blumei* and *Impatiens sultani*, measuring the break strength of the abscission zone over 6 days and 1 day, respectively.

Small cylinders of ethylene gas can be obtained from Matheson Gas Products (USA) and from BOC Special Gases (UK) (see Part C2). To fill a syringe from such a cylinder, a 50-cm-long plastic tube must be attached to the outlet of the cylinder regulating valve. The end of the tube must be closed with a Suba-Seal (from William Freeman & Co., see Part C2) or a rubber bung holding a coarse syringe needle pointing into the tube. By allowing ethylene to flush the tube for about 1 min, the air in it will be replaced by ethylene and a syringe can then be attached to the needle and filled with ethylene. The regulator valve is open while flushing but closed when filling the syringe. *Both operations must be carried out in a fume cupboard.*

If the quantity of ethylene to be injected into the experimental boxes to give 62.5 mg m⁻³ (50 p.p.m.) is too small, a dilution can be prepared by injecting 1 cm³ ethylene into 1000 cm³ of air in a bottle to give 12 500 mg m⁻³ (10 000 p.p.m.). A quantity of this diluted mixture can now be injected to result in the correct concentration in the sandwich boxes.

The concentration of IAA required for treatment of the explants in the third box, 10^{-1} mol m⁻³ (100 μM), is prepared by taking a 10 mol m⁻³ (10 mM) IAA solution in ethanol and adding 1 cm³ of this solution to 100 cm³ of melted *anhydrous* lanolin at about 60 °C. Thorough stirring and mixing is required during which the ethanol will evaporate. When solidified, the mixture is best applied as in Experiment 10.8.

French bean plants (*Phaseolus vulgaris*) are grown, twenty to a tray, for about 3 weeks to a stage in which the primary leaves are fully developed and the first trifoliate leaf has just begun to expand. Explants are prepared by cutting off the petioles of both primary simple leaves (see Fig. B10.7a) 2 cm up from their point of insertion in the stem and cutting the stem 1.5 cm below and above the point of insertion of the primary leaves. The explants are placed on the surface of a gel of 2 g agar dissolved in 100 cm³ water covering the bottom of a plastic sandwich box to a depth of 1.0 cm. The lids of these boxes can be made gas-tight by a band of Sellotape or autoclave tape. A small hole should be made in the lid with a hot needle so that a syringe needle can be inserted through it.

Ethylene gas is injected into one sealed box so that the concentration of the gas inside the box is 62.5 mg m⁻³ (50 p.p.m.). After the gas injection, the small hole in the lid must be sealed with tape. A second box prepared in the same way is left without ethylene, or better with ethylene-absorbing Purafil (from E. S. West, see Part C2) kept in a small perforated tube in the box (cf. Expt 11.13). A third box can be prepared with 62.5 mg m⁻³ (50 p.p.m.) ethylene and some of the petiole stumps of the explants treated with lanolin plus 10^{-1} mol m⁻³ (100 μM) indole-3-acetic acid (IAA) or with lanolin alone (cf. Expt 10.8). Finally, a fourth box could be provided which is ethylene-free, but with the explants treated as in the third box.

It is recommended to follow the time course of abscission by estimating the 'break strength' of the abscission zone. A series of weights from 50 to 300 g or plastic bottles filled with water to give an equivalent set of weights, each provided with a loop of thread, are required. Petioles of explants can then be held horizontally through the loops and gently lifted to raise the weights. The loop should be placed over the petiole stump 0.5 cm from its point of insertion. The break strength is the smallest weight that just snaps off the petiole. Average values for several petioles at different times during the various treatments should be

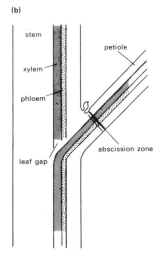

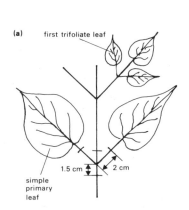

Figure B10.7 Illustration of (a) the method of preparing explants for Experiment 10.7 and (b) the position of the abscission zone near the petiole base (only half the stem is shown).

Time-course determinations with or without ethylene and with and without IAA will require daily measurements of break strength for 4 days. If a detailed time course of the ethylene treatments is desired, measurements of break strength will be needed every 12 h for 6 days — a lag period will become apparent. In the ethylene treatments, it is possible to delay auxin additions until 12, 24 or 36 h after administering the ethylene. During the first phase of abscission (up to approximately 20 h), auxin will inhibit the process; no matter when it is applied thereafter it is ineffective.

Ethylene treatment of flowers of *Rosa* and *Geranium* will cause petal abscission and that of *Nicotiana* and *Lycopersicum* whole flower abscission within 6–24 h. Green tomato and unripe raspberry fruit will show remarkably accelerated ripening when similarly treated.

determined. An initial lag time will be found during which there is no decline in break strength. This is followed by a progressive decline in break strength.

RELEVANT ANATOMICAL STUDIES

For observations of the anatomy of the abscission zone, *Coleus blumei* petioles and stems should be cut longitudinally followed by the preparation of thin hand-cut sections (approximately 0.005 cm) across the region of insertion of the petiole in the stem. Mounted in water or liquid paraffin, such sections will show the band of narrow specialised cells about three cells wide situated at about six to eight cortical cells along the petiole (see Fig. B10.7b). Staining good sections with ruthenium red (see Part C1) may show the middle lamellae which will be dissolved away during abscission (cf. Expt 10.8). For the best observations of middle lamellae stained with ruthenium red, collenchyma tissue found in ribbed stems and petioles is most suitable.

For the observation of the separation process at the middle lamellae between cells of the abscission zone, leaf blades of busy Lizzy (*Impatiens sultani*) should be cut off 36, 24 and 12 h before the practical class, leaving the petiole stumps on the plant. During the class, longitudinal sections of the base of the stumps near their points of insertion will show gaps between intact rounded cells of the fracture zone where the middle lamellae have been degraded by cellulase (cf. Expt 2.5). It will be noted that cells of the cortex and epidermis tend to separate, but not the xylem vessels in which tyloses may be observed.

10.8 Abscission

Based on information given by Dr K. Hardwick, University of Liverpool
cf. Expts 2.4 anatomy, 2.5, 10.7 & 10.9

Discs, 0.5 cm diameter, impregnated with 0.1 cm³ of stock solutions containing 10 mg 1000 cm⁻³ and 1 mg 1000 cm⁻³ indole-3-acetic acid (IAA) or naphthalene acetic acid (NAA) are required. Both are recommended for use so that a comparison of the effectiveness of the naturally occurring IAA and the synthetic NAA at equal concentrations can be made. The length of time required for abscission to occur in the NAA-treated plants will be greater than in those treated with IAA as NAA is not enzyme-degraded.

After impregnating the discs, they are allowed to dry and can be stored in Petri dishes in the dark in a refrigerator for no more than 2 weeks. Do not omit blank discs, impregnated with 0.1 cm³ water.

Stem sections, 2 cm long, of *Coleus blumei* plants with oppositely inserted petioles are cut so that the petioles are positioned in the centre of the nodal section. With a sharp blade the trimmed sections are further split longitudinally and the axillary buds as well as the leaf laminae excised. Each half node is placed immediately with its cut longitudinal surface on wet filter paper lying on a microscope slide. The transversely cut petiolar stumps (0.5–1.0 cm long), pointing upwards, are now treated with a drop of water followed by the application of dry prepared filter paper discs impregnated either with auxins or water (blanks). Alternatively, prepared gelatine capsules can be applied to the cut petiole. The slides carrying the treated material are placed in plastic boxes lined with moist filter paper, and kept at 20–25 °C. Abscission will be delayed in the material treated with auxin.

Intact plants should also be used; in this case the only practical method is that of attaching gelatine capsules to the cut petiole stumps, but they must be applied immediately after removal of the laminae to prevent the stumps from drying out. An additional experiment is recommended to study the effects of removing different proportions of the lamina, for

Empty gelatine capsules, No. 0, can be obtained from drug companies (e.g. Parke Davies, see Part C2). The rounded end must be pierced with a stout pin before the auxin, mixed with lanolin, is filled in with a small spatula, about 0.25 cm³ per capsule.

The lanolin is prepared by melting at 35 °C. Ten milligrammes IAA or NAA dissolved in 2 cm³ acetone are mixed with 100 cm³ melted lanolin and allowed to solidify before use. Do not omit capsules with plain lanolin (cf. Expt 10.7).

Application of capsules must be gentle, and they should enclose not more than about half of the petiole stump. It is important to ensure that air pockets are not trapped between petiole and lanolin preventing good contact between the two (Fig. B10.8).

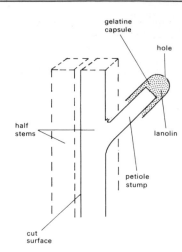

Figure B10.8 Longitudinal section of gelatine capsule filled with lanolin in position on petiole stump.

example 25, 50 or 75%. This can be done either at right angles to the mid-vein or parallel to it. Abscission may be found to be delayed if as little as 6% of the lamina remains. (cf. Relevant anatomical studies, Expt 10.7.)

10.9   Apical dominance
Based on information given by Dr K. Hardwick, University of Liverpool
cf. Expts 10.7 & 10.8

Similar sized, 3- to 4-week-old runner bean (*Phaseolus vulgaris*) plants with their two primary leaves and an apical bud developing about 6 cm above the primary leaf node are used in groups of three growing in a 15-cm-diameter pot at 20–25 °C. One plant is left intact. From the second and third, the apical buds are removed with a sharp blade so that the lengths of the shoots of the two plants are equal. Gelatine capsules loaded with either plain lanolin or a lanolin–hormone mixture (see Expts 10.7 & 10.8) are now applied to these plants.

Lengths and stages of development of axillary buds and the lengths of stems above the primary leaf node are recorded daily. Plants may need support by staking. Replication by student groups is needed and the mechanism of hormone intake by the plants should be discussed. Axillary bud growth will be suppressed in the plants that have been treated with the lanolin–hormone mixture.

10.10   De-repression of dwarf characters by gibberellic acid (GA₃)
Based on information given by Dr K. Hardwick, University of Liverpool
cf. Expts 2.1, 2.2, 10.1, 10.2 & 10.5

Growing plants for this experiment should be standardised by soaking seeds for 12 h in running tap water and planting in six rows of six seeds in trays (35 × 23 × 6 cm) with a good growing compost (e.g. a 50 : 50 mixture of Vitax, obtainable from Steetley Minerals Ltd (see Part C2), and John Innes No. 1 compost) at 20 °C. Temperatures above 25 °C must be avoided. After 14 days plants should be about 6–8 cm high.

Potted *dwarf* pea plants (*Pisum sativum* var. Meteor) are treated with drops of water and 10^{-2}, 10^{-1}, 1, 10 or 10^2 μg GA₃ (see below) placed in the axils of leaves of node 4 counted from the base up (see Fig. B10.10). Measurements of internode lengths and total plant height are recorded every 2 days and presented graphically. Best results are obtained if the treated node is at a stage of about 50% of full expansion.

Plants undergoing internode extension should be *loosely* supported on stakes.

Some trays should be planted randomly and the experimental merits discussed of applying treatments either randomly to randomly spaced plants, or methodically, one treatment for the neatly spaced plants in one row.

GIBBERELLIC ACID

Gibberellic acid stock solution contains 10 g GA_3 1000 cm^{-3}. The GA_3 is first dissolved in a few cubic centimetres of acetone and then made up to 1000 cm^3. Dilutions should be prepared so that 0.1 cm^3 added at a node is the desired dosage. Droplets should be applied with 1-cm^3 disposable syringes with fine needles.

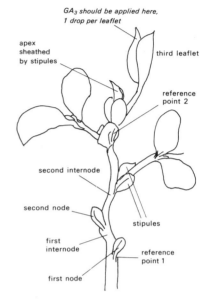

Figure B10.10 Diagrammatic representation of dwarf pea plant shoot showing positions of GA_3 applications.

Experiments involving the use of radioactive material must be carried out according to the safety regulations in force. All radioactive waste, including plant material, must be disposed of immediately after the conclusion of experiments.

10.11 Polar transport of auxin
Based on information given by Dr C. C. McCready, University of Oxford

Sections of opposite petioles of the primary leaves of dwarf beans (*Phaseolus vulgaris*) are used for the study of polarity and velocity of auxin transport.

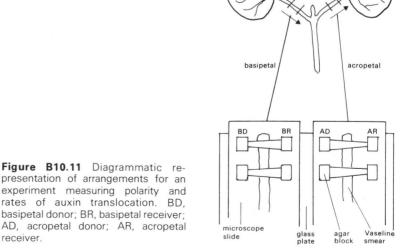

Figure B10.11 Diagrammatic representation of arrangements for an experiment measuring polarity and rates of auxin translocation. BD, basipetal donor; BR, basipetal receiver; AD, acropetal donor; AR, acropetal receiver.

The preparation of the agar blocks for 10 groups of students is accomplished as follows: 0.3 g agar is dissolved in 20 cm³ of water, and [1 − ¹⁴C]IAA (from Amersham International, see Part C2) is added to the hot sol to give a final concentration of 5×10^{-3} mol m⁻³ (5 μM). After mixing, the agar sol is drawn into 30-cm-lengths of 0.3-cm-bore capillary glass tubing plugged at one end with non-absorbent cotton wool. After chilling to gel the agar, a cylinder of the gel is extruded by pushing the cotton wool plug along the tube. The gel is sliced into 0.3-cm-long donor blocks with a multiple bladed cutter composed of spaced razor blades and the blocks are expelled from the cutter by means of a comb. Receiver blocks are prepared in the same way from plain agar gel without IAA. Inhibitors such as naphthylphthalamic acid, 0.5×10^{-3} mol m⁻³ (0.5 μM), fluorenolcarboxylic acid, 5×10^{-3} mol m⁻³ (5 μM) or tri-iodobenzoic acid, 5×10^{-3} mol m⁻³ (5 μM), may be incorporated in donor and receiver blocks, which are stored, for not more than a few hours, on glass slides in Petri dishes lined with damp filter paper. The dishes should be kept closed as much as possible to minimise drying of blocks and sections.

As an alternative to IAA, 2,4-dichlorophenoxyacetic acid may be used; because of its lower velocity of transport, overnight runs are convenient (McCready 1963).

To set up the experiment, a median line of Vaseline is extruded from a syringe along a 7.5 × 2.5 cm microscope slide in a Petri dish lined with damp filter paper. A row of 10 donor blocks is arranged down one side, and a row of 10 receiver blocks down the other. The side with the donor blocks should be marked to avoid confusion. The sections of petioles are laid across the line of Vaseline, and without delay the blocks are pushed into contact with the ends of the sections using the colour-coded glass rods. When placing the petiole sections on the line of Vaseline, great care must be taken to position them the correct way round.

As shown in Figure B10.11, ten paired petiole sections are placed between donor and receiver blocks of agar (i.e. with and without indole-3-acetic acid (IAA)). The glass plate carrying the complete experimental arrangement is kept in a large plastic box lined with moist filter paper at 25 °C. In addition to the box containing plant material, 10 blank sets each of donor and receiver agar blocks must be provided in a separate box to serve as standard donor and background receiver. At the end of the experimental time the agar blocks from each treatment and the blanks are analysed by scintillation counting.

The seedlings are grown in the glasshouse until petioles of the primary leaves are 1.0–2.5 cm long (11–14 days, depending on conditions). Petiole sections of 0.5 cm length must be cut with double blades set at a fixed distance in a block handle. Sections should be handled with forceps without crushing the tissue. Acropetal and basipetal ends should be marked.

The agar blocks must also be handled with forceps and there must be separate ones for each kind. Plastic forceps are very suitable for this. Three sets are needed and they should be colour-coded. When forceps are used in different treatments they should be thoroughly washed between treatments. Colour-coded tapered glass rods are useful for manipulating the agar blocks.

Possible treatments include:

(a) A comparison of acropetal and basipetal transport, run for 3 h. This will demonstrate polarity.

(b) Basipetal transport run for 1.5, 2.5, 3.5 and 4.5 h. This will give a time course of accumulation in receiver blocks from which the linear velocity of transport can be estimated (McCready & Jacobs 1963).

(c) A comparison of IAA alone and IAA plus an inhibitor. This will show the different effects of inhibitors on acropetal and basipetal transport (McCready 1968).

All treatments should be commenced at a known starting time. Because it takes time to set up the different treatments, it is suggested to set up acropetal and basipetal arrangements alternately. As the nominal time of the beginning of the treatment it is best to record the mean of the times when the first and last donor blocks were put in contact with the petiole sections.

The Petri dishes containing the slides bearing the petiole segments and the agar blocks are kept at 25 °C in an incubator for the experimental time. Pooled sets of 10 basipetal or acropetal receiver blocks are then transferred to separate liquid scintillation vials, each containing 10 cm³ of a water-tolerant scintillant, for assay of ¹⁴C. A spare set of 10 unused receiver blocks is similarly assayed to give the background count. All counts must be corrected for background radiation and, if necessary, for 'lost' agar blocks. If donor agar block counts differ by more than 5%, a correction for this will be needed. Acropetal transport and inhibitor effect are expressed as percentages of basipetal transport.

REFERENCES

McCready, C. C. 1963. *New Phytol.* **62**, 3–18.
McCready, C. C. 1968. In *Biochemistry and physiology of plant growth substances*, F. Wightman and G. Setterfield (eds), 1005–23. Ottawa: Runge Press.
McCready, C. C. and W. P. Jacobs 1963. *New Phytol.* **62**, 19–34.

10.12 Phytochrome and germination
Based on information given by Dr J. Hannay, Imperial College of Science and Technology

All seed supplies vary and preliminary tests must be carried out. Dark germination scores of up to 40% are still acceptable, because light scores should be about 90%. Usually scores can be recorded satisfactorily 3 days after the treatments were given. However, only seed that has produced a radicle of 1 mm in length or more should be counted. Shed testas and ungerminated seed are easily confused and 'total' counts can therefore add up to more than 50. Hence the need to count *exactly* 50 seeds.

It is best to count the seeds into the upturned lid, prepare the dish with filter papers and water, remove any trapped air bubbles and then scatter the seeds from the lid evenly on the filter paper so that they will all begin imbibition at the same time.

Phacelia does not respond to short light 'breaks', but 24 h illumination followed by darkness is still inhibitory.

The material originally used for this experiment was 'Grand Rapids' lettuce seed; this is now difficult to obtain and old seed is not reliable. A few of the newer varieties now available, such as the winter forcing lettuce 'Dandie' (from E. W. King & Co. Ltd or NSDO (see Part C2)), work well also. If such seeds are stored dry at between 3 and 7 °C in a sealed container, they will remain useful for at least 2 years. All seeds must first imbibe water in complete darkness for at least 1 h at room temperature before experimental treatments can be started. Exactly 50 seeds are evenly spread on two layers of 9-cm-diameter filter paper (Whatman No. 1) in a Petri dish containing 5 cm³ water. Clumping together of seeds is to be avoided. Petri dishes should be placed as shown in Figure B10.12 to ensure light-tightness and for ease in applying light treatments.

(a)   For a comparison, seeds of *Phacelia tanacetifolia* (Hydrophyllaceae), an uncommon garden plant, should be used in the introductory experiment. Two pairs of dishes, one seeded with lettuce and the other with *Phacelia*, are prepared. One dish from each set is treated in continuous light, the other in continuous darkness. When germination scores are recorded, it will be seen that the seeds respond in opposite ways to continuous white light.

(b)   For experimentation on light *quantity* lettuce seed only should be used, with a standard non-fluorescent light source which can be quite simple, e.g. a 100 W incandescent bulb at a distance of 1 m. Light treatments can be for 1 s, 1 min, 1 h and continuous. Results will show that the action of light is in the nature of a trigger. If desired, time and intensity of illumination can be adjusted so that the light supply remains constant at a level that will result in about 50% germination. It is necessary to choose a light supply resulting in less than maximum response if the aim is to detect any differences in response to light supply. It would be best to experiment with the following treatments and to interpolate to find the combination that will result in 50% germination:

$$0.5 \ \mu\text{mol m}^{-2}\text{ s}^{-1} \times 1, 10, 100 \text{ and } 300 \text{ s} \qquad (0.5 \ \mu\text{E m}^{-2}\text{ s}^{-1})$$
$$5.0 \ \mu\text{mol m}^{-2}\text{ s}^{-1} \times 1, 10, 30 \text{ and } 100 \text{ s} \qquad (5.0 \ \mu\text{E m}^{-2}\text{ s}^{-1})$$
$$50.0 \ \mu\text{mol m}^{-2}\text{ s}^{-1} \times 1, 3 \text{ and } 10 \text{ s} \qquad (50.0 \ \mu\text{E m}^{-2}\text{ s}^{-1})$$

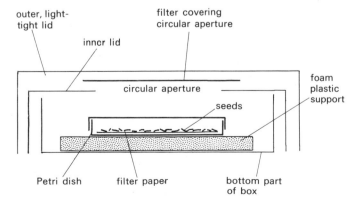

Figure B10.12 Longitudinal section of light filter box for exposing seeds to light of different wavelengths.

If the ecological significance of the seed properties is of interest, one treatment should be of white light filtered through a large green attached leaf, e.g. tobacco. Comparison treatments would be with red and far-red light as well as with leaf-filtered white light that has also passed through 2 cm of a solution of 1 g $CuSO_4$ in 100 cm³ water placed in a dish *above* the green leaf – this solution will absorb practically all the far-red light.

(c) Experiments on light *quality* require reasonably good, broad-band filters in the blue, green, red and far-red regions of the spectrum (from Rank Strand Electric, see Part C2). The filters must be 'built into' special boxes as shown in Figure B10.12. Treatments should be for 5 min in each wave band, and in white light and darkness. The light supply will differ in the different wave bands but this does not invalidate the results, because it has been shown that the light effect consists of triggering off some process. Percentage germination scores show that red light is as effective as white, but that the other wavelengths do not give clear-cut results, although blue and far-red seem to have an inhibitory effect.

(d) To test for an inhibitory effect, the following treatments should be given:

>complete darkness
>5 min red
>5 min red, followed by 5 min far-red
>5 min far-red, followed by 5 min red
>5 min red, 5 min far-red, 5 min red
>5 min far-red, 5 min red, 5 min far-red

If 5 min far-red is found ineffective in reversing 5 min red, an increase to 10 min far-red is recommended.

Effects of GA_3, kinetin, and ABA in combination with the light treatments can be studied at zero concentration of these substances and at the following concentrations added to the 5 cm³ of water in the Petri dishes: 100 mg 1000 cm⁻³ GA_3; 10 mg 1000 cm⁻³ kinetin; 3 mg 1000 cm⁻³ ABA. Incubation at 25 °C in continuous light and continuous darkness will show that GA_3 and kinetin stimulate germination in darkness, whereas ABA suppresses it in light.

Experiments can also be carried out at different temperatures, e.g. 20, 25 and 30 °C. It will be found that the light response is critically dependent on temperature.

The reversible effect of red and far-red will be obvious from percentage germination scores, as will be the fact that only the last treatment given is finally effective in determining percentage germination.

In order to maintain constant temperature during the different treatments, an incubator with a glass door will be convenient if a constant temperature room is not available. For light quality treatments, the outer lid of the special box may be removed because the 'built-in' filters already cover the central hole and the seeds can be illuminated for 5 min by the standard light source. Careful removal from the incubator for the duration of the light treatment is acceptable from the point of view of temperature control, except for the continuous light treatment. This can be given through the closed glass door of the incubator from the 100 W bulb suspended outside but close to the door.

11 Growth

11.1 Imbibition and germination
Based on information given by Professor E. W. Simon, Queen's University, Belfast
cf. Expts 1.4 & 11.13

Unless sterilised or dressed seeds are used, contamination will interfere with the results if the seeds are kept for longer than 3–4 days in water.

After measuring the volumes of batches of 25 dry pea seeds by immersion in 30 cm³ water contained in a measuring cylinder, one batch is kept for 3–4 days in 100 cm³ water contained in a 250 cm³ beaker and another in 100 cm³ of a 2×10^3 mol m^{-3} (2.0 M) NaCl solution. Likewise, 10 seeds are kept in a Petri dish on three filter paper discs and 10 cm³ water and another 10 seeds in a dish with papers wetted with 10 cm³ of 2×10^3 mol m^{-3} (2.0 M) NaCl. The papers must be kept moist by replenishing the liquids, and all treatments should be carried out at a uniform temperature of 20 °C.

Results correlate the volume increases after imbibition and osmosis in water and in 2×10^3 mol m^{-3} (2.0 M) NaCl with the percentage germination in the two media. Percentage germination in NaCl should be expressed as a percentage of the germination in water.

11.2 Temperature, germination and acid phosphatase activity
Based on information given by Professor E. W. Simon, Queen's University, Belfast
cf. Expts 2.10 & 11.13 anatomy

(A) TEMPERATURE AND GERMINATION

Between 20 and 25 seeds of cucumber (*Cucumis sativus*) or white mustard (*Brassica alba*) are sown in 9-cm-diameter Petri dishes each containing three filter paper discs and 10 cm³ water. Dishes are kept at 2, 5, 10, 15, 20, 30, 35 and 40 °C and the papers must remain moist. After 7 days the lengths of primary roots are measured to the nearest 0.5 cm and the percentage germination is recorded for each treatment and species. Plots of both lengths and percentage germination against temperature are prepared.

(B) ACID PHOSPHATASE ACTIVITY

This determination of enzyme activity depends on the metabolism of p-nitrophenyl phosphate which, on being hydrolysed, yields free phosphate and p-nitrophenol; this is yellow in alkaline solution and therefore suitable for colorimetric estimation using a blue filter. Tube number 1 (Table B11.2) is used to zero the instrument.

Table B11.2 Reagent mixtures for the colorimetric estimation of free p-nitrophenol.

					Tube				
	1	2	3	4	5	6	7	8	9
p-nitrophenol stock (cm³)	0.0	0.1	0.2	0.3	0.4	0.6	0.8	1.0	1.4
water (cm³)	1.5	1.4	1.3	1.2	1.1	0.9	0.7	0.5	0.1
10^2 mol m⁻³ (0.1 M) NaOH (cm³)	5.0	5.0	5.0	5.0	5.0	5.0	5.0	5.0	5.0

A calibration curve is prepared by using a stock p-nitrophenol solution of 1 μmol cm⁻³ and plotting absorbance against μmol p-nitrophenol.

For the determination of enzyme activity in roots of seeds germinated between moist filter papers at 20 °C, eight tubes are prepared, each containing 1 cm³ of 1 mol m⁻³ (1.0 mM) citrate buffer (see below) at pH 4.5 and 0.5 cm³ of 3×10^2 mol m⁻³ (0.3 M) p-nitrophenyl phosphate. Tubes must be conditioned for 15 min at the experimental temperatures (2–40 °C). Apical root sections, 2 cm long are excised from germinated seeds and collected on wet filter paper before placing them in batches of ten into temperature-conditioned test-tubes prepared as shown in Table B11.2. These tubes are now incubated for exactly 30 min at their respective temperatures and after 30 min 5 cm³ of 10^2 mol m⁻³ (0.1 M) NaOH are added to each tube. The liquid is next poured into a colorimeter tube, leaving the root sections behind. Absorbance is measured and compared with a blank containing the same reagents but no p-nitrophenyl phosphate. From the calibration curve, absorbance can be converted into μmoles p-nitrophenol. A final graph can be constructed of the effect of temperature on acid phosphatase activity as reflected by the amount of p-nitrophenol found after each treatment.

The objective of these two experiments is to correlate percentages of germination and enzyme activity in *one* species at a range of temperatures. Measurements of percentage germination, lengths of roots produced and enzyme activity should be plotted against temperature to provide a basis for discussion.

CITRATE BUFFER

Citrate buffer, approximately 1 mol m⁻³ (1.0 mM) at pH 4.5, is prepared by adding 47 cm³ of 1 mol m⁻³ (1.0 mM) citric acid to 53 cm³ of 1 mol m⁻³ (1.0 mM) trisodium citrate (cf. Part C1).

11.3 Phosphate transport from roots to shoots

Information given by Professor E. W. Simon, Queen's University, Belfast, based on an experiment devised by Professor J. S. Pate, University of Western Australia, Perth
cf. Expts 2.10, 9.2, 9.5 & 11.13

Five- to six-week-old potted pea (*Pisum sativum*) seedlings, about 20 cm high, should be detopped to about 1 or 2 cm above soil level 3 h before the experiment which involves the collection of sap-exudate is begun.

Standard phosphate solutions can be prepared by diluting a stock solution of 34.4 mg KH_2PO_4 in 1000 cm³ distilled water; this solution contains 24 µg cm⁻³ $(PO_4)^{3-}$. Dilutions to obtain 18, 12, 6 and 3 µg cm⁻³ are recommended and of each standard 7 cm³ are used in each colorimeter tube, except one tube containing 7 cm³ distilled water. To each tube are added 1 cm³ of diluted Analar perchloric acid (1 cm³ acid plus 9 cm³ H_2O), 1 cm³ amidol and 0.5 cm³ of the ammonium molybdate solution. Immediately after adding the $(NH_4)_6Mo_7O_{24} \cdot 4H_2O$ the content of the tubes is mixed. After 10 min when the colour has developed, D is determined.

The sap is collected with a 1 cm³ syringe during a 1 h period. The collected sap is transferred into a small *weighed* flask and because only some of the plants will exude, a record of the number of plants that did exude should be kept so as to be able to calculate the weight of exudate in grammes per hour per plant.

The sap is diluted 100 times with distilled water and 7 cm³ of the *diluted* sap are added to 1 cm³ perchloric acid plus 1 cm³ amidol and 0.5 cm³ *ammonium molybdate* (8.3 g 100 cm⁻³). After thorough mixing, the solution is transferred to a colorimeter tube and its optical density (D) determined after 10 min. The concentration of phosphate in the diluted sap can be determined by reference to a calibration curve prepared using standard phosphate solutions.

11.4 Nitrogen metabolism during germination
Based on information given by Dr M. J. Wren, University of Leeds
cf. Expts 4.6, 4.8 & 4.9

The 'dry' seed (day zero) needs to be soaked in water overnight in a refrigerator. The testa can then be removed from about half the number of seeds and the axis dissected out.

If eight time-treatments are too many, five may suffice. Harvest on days 0, 1, 2, 3 and either 5, 6 or 7.

The preparation of the material is as in Experiment 11.5, but using mung beans (*Phaseolus aureus*) allowed to germinate over 7 days at 25 °C in the dark. Sowing should be staggered so that the different ages will be ready for extraction on the same day. The growing medium is coarse perlite which is first saturated with tap water (not distilled, not deionised), allowed to drain and then put into polythene bowls to a depth of 8 cm. After sowing, the beans should be covered with a thin layer of dry perlite and the bowls covered with polythene until shoots have emerged. Harvesting can be carried out by flooding the perlite and gently removing the seedlings, roots and all.

When samples are taken, the testa must be removed first and 20 seeds or seedlings of each age separated into embryo axes and cotyledons. Separate extracts of each are prepared by grinding 20 axes or 40 cotyledons of each age in a mortar with 50 cm³ ice-cold 10³ mol m⁻³ (1.0 M) NaCl in 10² mol m⁻³ (0.1 M) phosphate buffer (see Part C1) at pH 7.0 containing 0.5 g *insoluble* polyvinylpyrollidone. After standing for 15 min, the extracts are filtered or centrifuged and made up to 100 cm³. From this preparation samples are taken for protein determinations by the dye-binding method (cf. Expt 4.8) and results expressed on the basis of seedling axis and pair of cotyledons.

11.5 Carbohydrate metabolism during germination
Based on information given by Dr M. J. Wren, University of Leeds
cf. Expts 2.2, 2.3, 9.4, 11.1 & 11.2

(A) STARCH-DEGRADING ACTIVITY

Barley (*Hordeum vulgare*) caryopses are used for this experiment; sowing should be staggered so that the different ages will be ready for extraction on the same day. The caryopses are first treated for 20 min in a solution of 1 g hypochlorite in 100 cm³ water (standard Milton) and very well rinsed thereafter. They are then placed on wet filter paper in a tray about 5 cm deep enclosed in a polythene envelope and kept at 25 °C in the dark. Five caryopses of each age (1–7 days) are taken and the whole seedlings including the grain are ground in a mortar with 10 cm³ ice-cold citrate buffer (see below) at pH 5.0, followed by filtration or centrifuging. Two cubic centimetres of each filtrate, or supernatant, are boiled immediately to inactivate enzymes and then frozen at −20 °C for later analysis.

Another portion of the unboiled filtrates or supernatants is analysed for starch-degrading activity. The samples from the second to the seventh day treatments will require dilution by adding 4 cm³ citrate buffer to 1 cm³ of each extract. All tests must be carried out in duplicate, including the blanks.

Of each unboiled extract (some of them diluted), 0.1 cm³ is added to 5 cm³ soluble starch solution (see Part C2) and the solutions are mixed. For the blank, 0.1 cm³ citrate buffer is used in place of the extract. The mixtures are incubated at 30 °C for exactly 15 min, after which time 0.2 cm³ iodine reagent (see Part C2) is added to each tube, mixed and diluted with 5 cm³ water for testing in a colorimeter with a red filter or for absorbance at 600 nm (6000 Å) in a spectrophotometer. The colorimeter should be set at maximum reading with the red filter using 5 cm³ of the starch solution mixed immediately before use with 0.2 cm³ iodine reagent and 5.1 cm³ water. The spectrophotometer is first set to zero with 0.2 cm³ iodine reagent plus 10.1 cm³ water for absorbance at 600 nm (6000 Å).

A calibration graph is obtained with 5-cm³ samples of 0.05–0.2 g soluble starch in 1000 cm³ water added to 0.2 cm³ iodine reagent and 5.1 cm³ water.

Starch-degrading activity is expressed as milligrammes of soluble starch degraded per seed or seedling against age.

Citrate buffer

Citrate buffer, approximately 25 mol m⁻³ (25 mM), is prepared from two solutions A and B:

solution A is a stock solution of 4.80 g 1000 cm⁻³ citric acid;
solution B is a stock solution of 7.35 g 1000 cm⁻³ tri-sodium citrate·2H₂O.

A mixture of 400 cm³ solution A plus 600 cm³ solution B gives 1000 cm³ buffer at pH 5.0 (the pH should be checked and may need adjustment).

(B) TOTAL REDUCING SUGAR CONTENT (for details of reagents A–D, see below)

The samples of extracts that had been previously boiled and afterwards frozen are now thawed and diluted in the proportion 1 cm³ extract to 9 cm³ water.

From each treatment, duplicate material is prepared by mixing a 2.5 cm³ diluted sample with 2.5 cm³ reagent A, water replacing the sample in the blanks. All mixtures are placed in a boiling water-bath for 15 min and then cooled in tap water.

Using the blanks first, 1 cm³ reagent B is run gently down the side of the tube and when it reaches the extract, 1 cm³ reagent C is quickly added and well mixed in by shaking. After half a minute all precipitate will

---

The starch solution is first placed into the test-tubes, then extracts are added and, after incubation, the iodine reagent, treating the test-tubes in the same order as before.

Avoid air bubbles in the colorimeter and spectrophotometer tubes.

Zero readings with any of the samples mean that all starch has been degraded; higher dilutions will have to be made if the experiment is to be repeated.

Starch content per tube rather than starch concentration should be calculated and used in plotting the calibration graph. This makes the calculation of starch-degrading activity of the extracts simpler.

Before placing tubes with mixtures into the boiling water-bath, reagent A must be well mixed in with the extracts and, during the heating in the water-bath, tubes should be closed with a marble or aluminium foil.

Each tube should then be dealt with individually. After adding reagents B and C, the precipitate will clear and the solution plus several rinses should be titrated immediately.

have dissolved and the clear yellow liquid can be transferred into a 100 cm³ conical flask. All tubes must be well rinsed with water and these rinses added to the flasks.

Titration must be carried out immediately with reagent D, adding a drop or two of soluble starch solution when the titre becomes straw-coloured – the end-point under swirling is colourless. The difference in titration between blanks and samples indicates the reducing sugar content because 1 cm³ titration difference is equivalent to 0.0675 mg glucose.

Results are expressed as milligrammes glucose equivalent per seed or seedling.

REAGENTS

Reagent A
    28 g anhydrous disodium phosphate Na_2HPO_4
    100 cm³ 10^3 mol m⁻³ (1.0 M) NaOH
    40 g sodium potassium tartrate
    180 g anhydrous sodium sulphate Na_2SO_4
    0.892 g potassium iodate KIO_3 in 25 cm³ water
    8 g cupric sulphate $CuSO_4 \cdot 5H_2O$

    made up to 1000 cm³

Reagent B
    2.5 g potassium iodide in 100 cm³ water
    Add a knife tip of Na_2CO_3 to improve keeping qualities.

Reagent C
    10^3 mol m⁻³ (1.0 M) H_2SO_4

Reagent D
10^2 mol m⁻³ (0.1 M) sodium thiosulphate (24.8 g $Na_2S_2O_3 \cdot 5H_2O$ in 1000 cm³) is diluted to 2.5 mol m⁻³ (2.5 mM) not more than 1 day before use, adding 2 cm³ of 10 g NaOH in 100 cm³ water to each 1000 cm³ of diluted solution to protect against atmospheric CO_2.

(C)   IDENTIFICATION OF SUGARS

For this separation extracts are prepared from five germinated seeds or seedlings for each age (0-, 1-, 2- and 4-day material) by crushing in a mortar containing 5 cm³ water. After filtering, 2 cm³ of filtrate must be boiled to inactivate enzymes. Then 10 drops of each extract are gradually loaded as single spots on Whatman No. 1 chromatography paper together with 1 drop each of marker solutions of 10 g sucrose, maltose, fructose and glucose in 100 cm³ water. The spots should be loaded on to a pencil line 10 cm from the upper edge of the paper.

The descending chromatogram is developed in a solvent of 7 vol. ethylacetete, 2 vol. iso-propanol plus 1 vol. water for 36–48 h, allowing the solvent to drip off the paper in order to obtain good separation.

The inverted paper is hung up to dry and is then dipped through diphenylamine chromogenic reagent (see below) as horizontally as possible. Afterwards the paper must be placed on filter paper to remove

While heating in the oven at 100 °C, the paper must be watched and removed when spots appear before the background darkens.

One spot may not correspond to any of the marker spots – this is likely to be maltotriose if it is close to the starting line.

The chromogenic reagent is best kept in the shallow depression of a 19 × 19 × 4 cm glass brick. If evaporation is kept down, 220 cm³ will be sufficient for several chromatograms.

excess reagent and then, in the form of a loose roll standing upright in an oven, it is heated at 100 °C for about 5 min. As there is no solvent front, R_g values are calculated (the R_g value is the ratio of the distance travelled by the compound to the distance travelled by a glucose marker); R_g glucose $= 1.0$; R_g fructose >1.0; R_g sucrose <1.0.

DIPHENYLAMINE CHROMOGENIC REAGENT

2 g diphenylamine in 100 cm³ acetone	5 vol.
2 g aniline in 100 cm³ acetone	5 vol.
concentrated phosphoric acid	1 vol.

The reagent is prepared by mixing the above in a measuring cylinder immediately before use and stirring with a glass rod until the precipitate dissolves.

REFERENCE

Somogyi, M. 1945. *J. Biol. Chem.* **160**, 69–93.

11.6   Seed tolerance of anaerobic conditions
Based on information given by Professor E. W. Simon, Queen's University, Belfast
cf. Expts 7.3 & 11.7

Seeds of *Pisum sativum*, *Cucumis sativus* and *Oryza sativa* are placed, 12 of each species, in three pairs of conical flasks. One of each pair of flasks has the 12 seeds placed on a moist filter paper; the other has enough water in it to submerge the seeds. After 4–5 days percentage germination under aerobic and anaerobic conditions can be compared among species. All treatments should be at a uniform temperature of 20 °C.

Some seeds of each species will germinate under both conditions, but percentage germination and amount of growth made subsequently will differ from species to species. In addition, due to leakage of solutes from peas, the water will soon become contaminated with microorganisms and seedlings may deteriorate.

11.7   Tolerance of anoxia during germination
Based on information given by Professor R. M. M. Crawford and Dr A. M. Barclay, University of St Andrews
cf. Expts 7.3 & 11.6

Natural oxygen starvation occurs in most seeds during early stages of germination. An extension of this period due to waterlogged soil can cause serious reductions in percentage germination in some seeds, whereas others can endure anoxia for some time. Reductions in percentage germination also occur when seeds are suddenly exposed to unlimited water causing physical injury to the delicate membrane system. The percentage reduction in germination due to anoxia and the danger of physical injury to seeds are affected by prevailing temperatures. Experiments with different seeds under different treatments are instructive.

Pea (*Pisum sativum*), lettuce (*Lactuca sativa*), rice (*Oryza sativa*) and rye grass (*Lolium perenne*) are species in which tolerance of anoxia ranges from 24 h (pea) to 14 days (rye grass) at 20 °C.

Seeds should be soaked in tap water overnight and then well rinsed in distilled water. Twenty-five of these seeds are placed on two layers of Whatman No. 3 paper in 500-cm^3 Erlenmeyer flasks with a double bore rubber bung allowing for connection to an oxygen-free nitrogen cylinder. Some of the flasks are thoroughly flushed with nitrogen and sealed. Together with others containing ordinary air, they are incubated at 5, 10, 15, 20 and 25 °C, then the seeds are removed from their treatment flasks at 1-day intervals and planted in trays with moist, but not wet, sand. To be able to express the effects as a reduction in percentage germination, a comparison with control seeds is necessary and these must be planted out in parallel with the treated seeds. The greatest differential effect will be found at temperatures that the seeds would experience naturally.

Daily rates and final percentage germination are recorded and plotted as graphs of percentage germination (seedling emergence) *v.* length of anoxia treatment in hours for the different species; or for one species only (pea) *v.* anoxia treatment at different temperatures. From these graphs estimates can be arrived at for the time or temperature of anoxia treatment that would produce 50% reduction in seedling emergence for different species.

A further exercise can be carried out with pea seeds treated in anaerobic conditions for different lengths of time or at different temperatures as before, but determining the ethanol accumulation in the seeds and plotting against percentage emergence.

11.8 Heavy metal ions and plant growth
Based on information given by Professor E. W. Simon, Queen's University, Belfast.

Although traces of heavy metal ions are needed by most plants, in higher concentrations they are toxic for most except special heavy metal tolerant organisms. To demonstrate the effect, the fungus *Trichoderma viride* (from Commonwealth Mycological Laboratories, see Part C2) can be used.

Trichoderma viride is inoculated centrally into a plate formed by 45 g Czapek Dox agar (CM97; from Oxoid Ltd, see Part C2) in 1000 cm^3 water in a 9-cm-diameter Petri dish. After 2 days, samples are taken with a No. 3 cork-borer (0.4 cm internal diameter) from the outer periphery to where the fungus has spread. The discs of infected agar taken with the cork-borer are now centrally placed on sterile agar plates that contain 0, 10^{-3}, 10^{-2}, 10^{-1}, 1 and 10 mol m^{-3} (1 μM–10 mM) $CuSO_4$. After 1 week of incubation at 20 °C the spread of the fungus is measured at two diameters at right angles to each other and the average diameters are plotted against the logarithm of the $CuSO_4$ concentration.

11.9 Water stress and leaf extension growth
Based on information given by Dr W. J. Davies, University of Lancaster

Leaf extension growth, mainly by cell enlargement, can be measured reliably in a class experiment by the use of a piece of home-made equipment. The most suitable leaves are those of grasses and cereals. The leaf

After the anaerobic incubation in Erlenmeyer flasks, a small sample (five seeds) is killed immediately in liquid nitrogen and then transferred to a jar with 20 cm^3 6% perchloric acid for storage at −20 °C overnight. Proteins are removed by homogenising and centrifuging at 5000 *g* for 25 min at 0–5 °C followed by decanting the supernatant and neutralising the perchloric acid with 5×10^3 mol m^{-3} (5 M) K_2CO_3, using an aqueous phenolphthalein indicator. This will precipitate the protein when stored in a refrigerator for 2–3 h.

The final supernatant can now be analysed for ethanol with alcohol dehydrogenase as supplied in the test kit by Boehringer Co. (see Part C2).

Cork-borers used for sampling must be sterilised by dipping into absolute alcohol and then burning off with gentle heat; they may be used only after properly cooling to room temperature. When inoculating the test plates, raise the lid of the dish slightly to place the inoculum disc; avoid breathing on the agar plate or the sample disc.

The attention of students must be drawn to the implications of plotting the logarithms of concentrations (see Part A).

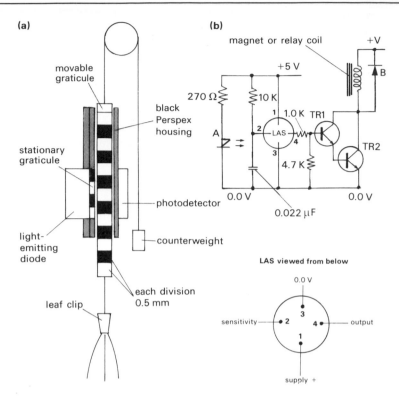

Figure B11.9 (a) Longitudinal section of apparatus used to measure extension growth of leaves. (b) Circuit diagram for the operation of the graticule apparatus measuring leaf extension growth. A, infra-red light-emitting diode (RS cat. no. 306–077); B, diode (RS cat. no. 1N 4001); LAS, light-activated switch (RS cat. no. 305–434); TR1, transistor (RS cat. no. BC 108); TR2, transistor (RS cat. no. BD 131) (see RS, Part C2).

tip is held in a Perspex clip attached to a thread which operates a graticule sliding past a matching stationary graticule. This allows a light to shine on a photodetector, as shown in Figure B11.9a.

The light source for the measuring device is an infra-red light-emitting diode operated from a 5 V d.c. supply (see Fig. B11.9b). A chart recorder with 10 V full-scale deflection is connected between output and negative rail. Readings could be taken at 30-min intervals if no recorder is available.

The graticules can be made from strips of photographic plate or film glued on to glass on to which the reduced image of a hand-made master graticule on Perspex (large scale) has been photographed. If no recorder is available, the graticule divisions must be numbered.

Leaf extensions up to 0.6 cm h^{-1} can be measured in well watered illuminated maize seedlings and this can be compared with the growth of water-stressed seedlings and others that have experienced water stress previously.

11.10   Measurement of thermonasty
Based on information given by Dr D. Idle, University of Birmingham

Thermonasty can best be studied by measuring the movement of petals in *Crocus ancryensis*, but flowers of other species and of different genera, for instance *Tulipa* (cut flowers), may also be used. Four plants are grown in 10-cm pots kept in a cold frame to flower early in spring. With the flower well formed, the potted plants can be preserved for 2 weeks in a

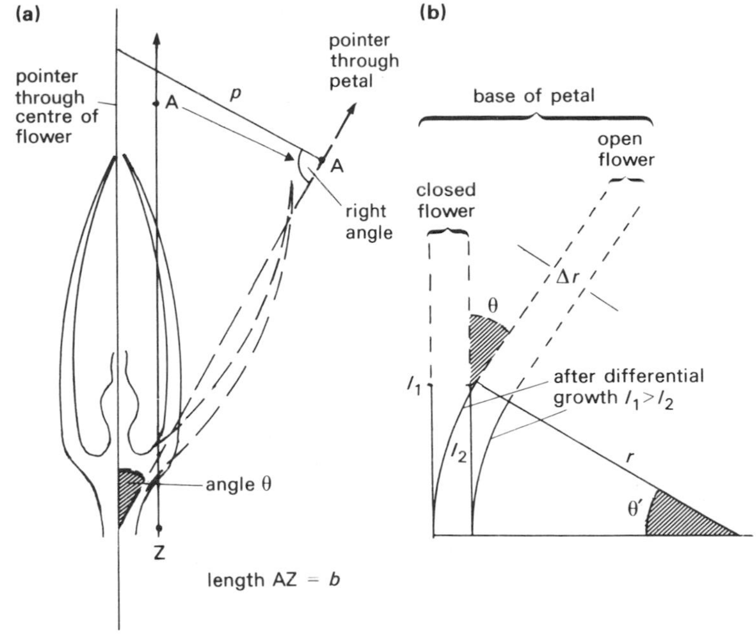

Figure B11.10 (a) Longitudinal section of crocus flower with the two pointers in position (full lines for the closed flower; dashed lines after opening). θ is the angle through which the petal moves; the fixed length of the pointer $AZ = b$; the increasing distance travelled by point A as the flower opens $= p$. (b) Enlarged longitudinal section of the base of the petal with sides l_1 and l_2 equal before differential growth takes place and after growth when $l_1 > l_2$ and angle θ is formed.

The pointer impaling the outer petal must not be pushed too far down and must not touch another petal; its end must remain free. Insertion of the pointer requires holding the petal gently between thumb and forefinger. The mark A should be fixed with marking ink.

cold room at between 0 and 1 °C. On transference to the laboratory at about 20 °C, flowers will begin to open within 10 min and attain full opening in about 1 h. Transfer back into the cold room will cause slow closure and the plants will be ready for re-use the following day.

Students need two glass pointers, 0.03 cm thick and 10 cm long. These pointers can be issued or, better, students can be shown how to draw them out from glass rod. One of the pointers is pushed vertically down to the centre of the closed flower, impaling style and receptacle; the other is gently positioned through an outer petal as shown in Figure B11.10a. The two pointers will be near parallel to begin with and as the flower opens they will gradually form an increasing angle.

The angle formed between the pointers can be calculated by measuring p shown in Figure B11.10a with a ruler, say every 10 or 15 min. The ruler must be held at mark A at right angles to the pointer going through the petal; b is constant and is the length of the petal pointer AZ in Figure B11.10a. The angle θ is determined from its tangent p/b (found in tables or from calculators) and the changes in θ are plotted against time.

Thermonasty in *Crocus* (not in all flowers) is the result of differential growth at the petal base. In order to estimate the degree of this growth, the thickness of the petal base needs to be measured from transverse sections of the base of the flower.

CALCULATION OF DIFFERENTIAL GROWTH AT PETAL BASE

Figure B11.10b shows the situation at the start of the experiment when the petal is straight and the two sides, l_1 and l_2 of its base are of equal length; after differential growth causing the flower to open by bending the petal outwards, l_1 will be longer than l_2 and the petal will have moved through angle θ. $\theta' = \theta$, since the angle between two tangents of a circle equals the angle between the radii to these tangents.

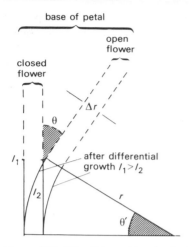

Figure B11.10b Enlarged longitudinal section of the base of the petal with sides l_1 and l_2 equal before differential growth takes place and after growth when $l_1 > l_2$ and angle θ is formed.

Three-day-old maize seedlings are suitable for geotropic experiments. Results may be observed after only 5 min and they will be clear after 20 min.

The pure agar medium is only suitable for short-term experiments lasting no longer than 8 h.

Figure B11.10b also shows the radius r of the circle through which the petal base has been bent and the radii enclosing the angle θ. The two unequal sides of the petal base l_1 and l_2 are shown as arcs in circles with radius $r + \Delta r$, the latter representing the separation of l_1 from l_2. If θ is measured in radians:

$$l_1 = \theta(r + \Delta r) \quad \text{and} \quad l_2 = \theta r$$

$$\text{Thus} \quad l_1 - l_2 = \theta(r + \Delta r) - \theta r$$

$$= \theta \Delta r$$

If r corresponds to the thickness of the petal, then the difference in growth of the two sides of the petal base needed to produce a curvature measured at the centre will equal $\theta \Delta r$. By definition, 2π radians = 360°; hence: $1° = \pi/180$ radians. Thus if θ is measured in degrees, the extra growth on one side $l_1 - l_2 = \pi/180\theta r$ where θ has been obtained from p/b and Δr is the thickness of the base of the petal.

11.11   Acid efflux patterns
By Professor H. Meidner, University of Stirling
cf. Expts 1.8 & 1.12

Proton extrusion is considered to increase the plasticity of cell walls and possibly change other properties of the cellulosic materials in the wall. Cell extension growth would thus be promoted. If proton extrusion is more pronounced on one side of a structure than on the other, curvature would result due to uneven wall extension. The following experiments seem to support these hypotheses as proton extrusion would seem to be more pronounced on the outside of the curvature.

The most suitable material for this exercise is provided by seedlings of maize (*Zea mays*) growing in a medium of 0.6 g non-nutrient agar in 100 cm³ water. The agar medium is adjusted to pH 5.0 with 10^2 mol m⁻³ (0.1 M) HCl and coloured with 5 cm³ of 0.7 mol m⁻³ (0.7 mM) bromocresol purple indicator, resulting in an orange colour. After boiling the agar sol, it is poured to a depth of about 0.4 cm into a 10-cm-diameter Petri dish, taking about 15 cm³ of the agar sol.

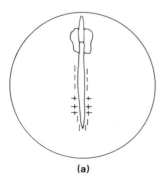
(a)

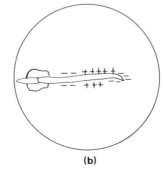
(b)

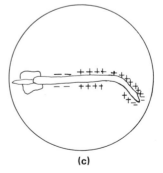
(c)

Figure B11.11 Pattern of pH changes in an agar medium containing bromocresol purple indicator dye surrounding the primary root of a germinating maize seed (*Zea mays*) undergoing geotropic curvature: (a) after 2 min; (b) after 20 min; (c) after 120 min.

Colour changes of the agar become clearer if the plates are viewed above a light box.

H⁺-efflux is known to be correlated with auxin concentration; the experiment described here is thus relevant to the Cholodny-Went hypothesis of tropic curvature. If auxin transport inhibitors are incorporated into the agar medium, the experiment can be developed further.

When the agar has cooled, seedlings are placed so that about one-half to two-thirds of the root surface is embedded in the agar. For the orientation of the seedlings in a vertical position, the Petri dishes are suitably placed as indicated in Figure B11.11. Geotropic curvature and its accompanying H⁺-efflux will be indicated by changes in colour of the agar medium. From the original orange (pH = 5.0), the colour will change to yellow where H⁺-efflux occurs and to red where OH⁻-efflux or H⁺-intake occurs. Colour changes become noticeable before any curvature can be seen. If watertight fingerprint ink marks are made on the root, the zones of elongation will become evident.

REFERENCE

Mulkey, T. J., K. M. Kuzmenoff and M. L. Evans 1981. *What's New in Plant Physiology*, **12**, 9–12. (These authors give further examples involving negative geotropism and phototropic curvature of shoots.)

11.12 Fundamentals of growth analysis
Based on information given by Dr R. Hunt, University of Sheffield

When setting up a growth analysis experiment with different species, allowance must be made for differences in the time taken for seeds to germinate.

Two soils of contrasted fertility and two light treatments (one shaded, one not) create four growing conditions involving variation in several environmental factors.

Each combination of species and growing conditions supplies one of each of the growth-analytical derivatives for each of its harvests or harvest intervals.

For class exercises in growth analysis, whether in controlled growth rooms, glasshouses or outdoors, statistically designed experiments are essential. A suitable design would be: two species × four growing conditions × four sequential destructive harvests, with as much replication as possible. Good measurement of environmental factors should be one of the students' tasks.

Plants must be 'washed' out of the soil in order to recover the whole root system. Separated into roots, leaves and stems or stem bases, plant materials are then blotted gently to remove surface moisture and weighed quickly to obtain fresh weights (*FW*), in milligrammes. Delays cause moisture loss.

Total leaf area per plant (L_A), in square millimetres must be determined next by any of the reliable methods available. After this, plant components are put into separate, labelled paper bags for drying at 95 °C to constant weight. This will be the dry weight (*W*).

(1) *FW/W* ratios can now be calculated; these will vary between species. As a rule, it will be found that they decrease with age of plant, increase with shading and decrease with nutrient deficiency.

(2) Root/shoot ratios (*R/S*) are expressed as $W_{root}/W_{total\ shoot}$. It will be seen that these ratios vary with species and that they generally decrease with age of plant and shade, but increase with nutrient deficiency.

(3) Other growth functions may next be calculated:

relative growth rate (*RGR*)	=	unit leaf rate (*ULR*)	×	leaf area ratio (*LAR*)
an expression of 'growth efficiency'		an expression of 'leaf efficiency'		an expression of 'leafiness'

Instantaneously, these are defined as:

$$\frac{1}{(W)} \cdot \frac{d(W)}{dt} = \frac{1}{(L_A)} \cdot \frac{d(W)}{dt} \times \frac{(L_A)}{(W)} \quad .$$

where $d(W)/dt$ represents the increase in W with unit time, t.

Each of these also has a mean value for specific intervals between harvests:

(a)   Mean (RGR) over a period $_1t$–$_2t$ in days is calculated as:

$$\overline{(RGR)} = \frac{\log_e {_2(W)} - \log_e {_1(W)}}{_2t - {_1t}} \text{ mg mg}^{-1} \text{ day}^{-1}$$

$_2(W)$ and $_1(W)$ refer to total plant dry weight at times $_1t$ and $_2t$; $_2(L_A)$ and $_1(L_A)$ refer to total leaf area at times $_1t$ and $_2t$.

(b)   Mean (ULR) = $\overline{(ULR)}$ = $\dfrac{_2(W) - {_1(W)}}{_2(L_A) - {_1(L_A)}}$

$$\times \frac{\log_e {_2(L_A)} - \log_e {_1(L_A)}}{_2t - {_1t}} \text{ mg mm}^{-2} \text{ day}^{-1}$$

(c)   Mean (LAR) = $\overline{(LAR)}$ = $\frac{1}{2} \left(\dfrac{_1(L_A)}{_1(W)} + \dfrac{_2(L_A)}{_2(W)} \right) \text{ mm}^2 \text{ mg}^{-1}$

(d)   There is one further relation that can be derived:

leaf area ratio (LAR)		specific leaf area (SLA)		leaf weight ratio (LWR)
'leafiness' (as before)	=	an expression of 'leaf density'	×	an expression of 'productive investment'

These may either be calculated directly, as:

$$\frac{(L_A)}{(W)_{\text{total plant}}} = \frac{(L_A)}{(W)_{\text{leaf}}} \times \frac{(W)_{\text{leaf}}}{(W)_{\text{total plant}}} \quad ,$$

or as mean values for specific intervals in the same manner as $\overline{(LAR)}$ shown under (c) above.

All terms can now be tabulated for the various intervals and relevant conclusions can be drawn. A specimen table layout is shown for an experiment on shading and nutrient stress (contrasted soils) with two contrasted species.

Consistent results have been obtained with two native grasses of different growth rates. If crop plants are to be used, differences are likely to be smaller. The number of harvests depends on the experimental design. In the example referred to here, there were 10 harvests of each of six

Specimen layout for results

Podzol (pH 4.2)				Garden soil (pH 6.0)			
Full light		50% shade		Full light		50% shade	
Species 1	Species 2	Species 1	Species 2	Species 1	Species 2	Species 1	Species 2

• $_1(W)$ and $_1(L_A)$

$_2(W)$ and $_2(L_A)$

$_{1-2}(\overline{RGR})$

$_{1-2}(\overline{ULR})$

plus ratios

and so on, for intervals 2–3, etc.

individual plants, twice weekly for 2 weeks. This required 240 60-mm diameter pots with two plants in each, all occupying 1 m² of bench space.

REFERENCE

Hunt, R. 1978. *Plant growth analysis*, Ch. 3. Studies in biology, No. 96. London: Edward Arnold.

11.13 The Salford auxanometer for roots
Based on information given by Mr P. St J. Edwards, University of Salford cf. Expts 11.1–11.3 & 11.9

Seeds of maize (*Zea mays*), lupin (*Lupinus alba*) and runner bean (*Phaseolus vulgaris*) are recommended, as they produce reasonably straight primary roots. Pea (*Pisum sativum*) can be used but its root growth tends to be irregular, being very sensitive to slight temperature variations. The auxanometer is shown in Figure B11.13a.

If studies are made with different gases in the atmosphere surrounding the seedling, the gas is injected to give the desired concentration, e.g. 2×10^3 mg m^{-3} CO_2 (1000 p.p.m.). An absence of ethylene can be arranged for by suspending a perforated tube containing Purafil (from E. S. West, see Part C2) inside the auxanometer (see Fig. B11.13). If the effects of different light qualities are to be studied, the auxanometer can be enclosed in different Cinemoid filter sleeves (from Rank Strand Electric Co.; see Part C2).

The apparatus must be kept vertical and positioned firmly in a low clamping position on a bench as vibration-free as possible in order to avoid the meniscus touching the root tip prematurely. There should be

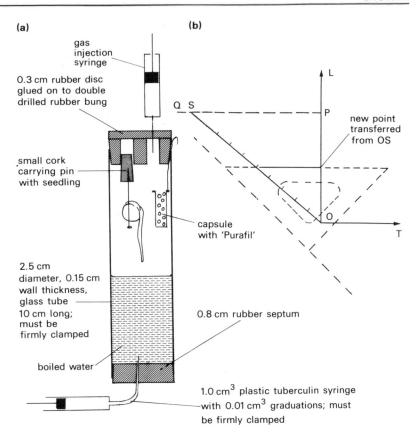

Figure B11.13 (a) The Salford auxanometer. The Pyrex barrel as well as the 1 cm³ graduated syringe must be securely clamped and the whole apparatus fairly vibration-free. All air bubbles must be expelled from the syringe. The apparatus can be enclosed in colour filters (see Part A) to test the effect of different wavelengths on root growth. (b) Graphical correction of plotted results obtained with the Salford auxanometer. Point P corresponds to a whole number of millimetres of meniscus movement, e.g. 0.1 cm corresponding to 0.65 cm³. PQ is then drawn at right angles to OL and point S chosen so that OS can be subdivided into whole numbers of centimetres. If P has been plotted on the y-axis at 6.5 cm then OS should be drawn with a length of 10 cm. To transfer graduations from OS to OL a set square and ruler are used as shown.

(a)

gas injection syringe

0.3 cm rubber disc glued on to double drilled rubber bung

small cork carrying pin with seedling

capsule with 'Purafil'

2.5 cm diameter, 0.15 cm wall thickness, glass tube 10 cm long; must be firmly clamped

0.8 cm rubber septum

boiled water

1.0 cm³ plastic tuberculin syringe with 0.01 cm³ graduations; must be firmly clamped

(b)

new point transferred from OS

Uniform behaviour of the meniscus inside the auxanometer tube can be achieved by scouring the tube with a bottle brush and a hot solution of 5 g Na_2CO_3 in 100 cm³ water, followed by copious rinsing under a cold water tap. Detergents are not recommended as they may contaminate the rubber bungs and then hinder root growth.

no air bubbles in either the auxanometer or the syringe (use distilled or boiled water). The apparatus requires calibration by pipetting 50 cm³ water into the tube and measuring the rise of the resulting water level; from this, the rise of the column in millimetres corresponding to the movement of the syringe plunger by one scale unit can be determined and experimentally verified.

All treatments must be replicated on account of variability. To obtain suitable seedlings, it is advisable to soak seeds in running or shallow water overnight and then sow these with the radicles downwards into dibbled holes in moist, sharp sand contained in boxes from which the seedlings can be removed with ease. After sowing, the seeds should be covered by a thin layer of dry sand. Boxes should be watered by a fine spray and covered with polythene. Timing for the different species will vary, but seedlings with a radicle of about 1.5 cm in length should be available for the start of the experimental measurements. Removal and impalement of the seedlings on pins without in any way damaging the radicles must be gentle and quick. Each pinned seedling is then inserted in the auxanometer with the root *vertical* and its tip as central as possible.

Setting the auxanometer meniscus can be carried out in three stages: filling to below the radicle tip from a wash bottle, adjusting the seedling on the pin to within 0.2 cm of the water level and finally moving the upper rubber bung slightly so that the root tip just touches the meniscus. When the syringe plunger is moved backwards by about 0.5 cm, the meniscus should break clear of the root tip.

Constant temperature control is desirable. Readings should be related to temperature measurements — a thermocouple (see Part A) could be incorporated in the auxanometer tube.

As there are some small sources of error that do not cancel each other, it is appropriate to fit a straight line *under* the majority of points plotted rather than draw a line through the middle of the distribution of points. The main cause of error is the premature connection of the meniscus due to vibration, resulting in excessively high points. The less common error is 'overshoot', i.e. failure to halt the plunger in the syringe at the correct point, resulting in low points on the graph.

It is recommended to record syringe units in fractions of cubic centimetres on the *y*-axis between readings. If the diameter of the tube of the auxanometer is as indicated in Figure B11.13a, the calibration will be (within 2%) 2 mm of meniscus movement corresponding to 1 cm³ of the syringe scale. For tubing of different diameter, a calibration can subsequently be fitted geometrically as shown in Figure B11.13b in which the syringe units in cubic centimetres had originally been plotted on the ordinate. Next, a point P on the ordinate is chosen that corresponds to a whole number of millimetres of the meniscus movement, and line PQ is drawn at right angles to OL. Point S is then chosen so that line OS is longer than OP and can be subdivided into whole numbers of centimetres. For instance, if OP reads as 0.65 cm³, corresponding to 0.1 cm of meniscus movement, and has been plotted at 6.5 cm on OL, then OS should be 10 cm long and subdivided into centimetres.

By use of a set square and ruler, graduations can be transferred to OL as shown in Figure B11.13b.

Measurements of radicle growth are carried out at 1- to 2-min intervals. At the start, the meniscus will not be in contact with the radicle. By pushing the syringe plunger gently in, a position will be reached when the meniscus 'blips' on to the root tip; the exact time and syringe scale reading must be recorded. This procedure is repeated at 1- to 2-min intervals and each reading in cubic centimetres is plotted against time. Points will not necessarily lie on a straight line, as errors, chiefly due to vibration and to overshoot by the syringe plunger, are not uncommon. A graphical correction is recommended as shown in Figure B11.13b.

RELEVANT ANATOMICAL STUDIES

CELL PRODUCTION

Ten maize seeds which have germinated in a Petri dish on moist filter paper are used to measure the average length of radicles and the average increase in length over 48 h. This is followed by cutting, by hand, longitudinal sections of the root where it has a diameter of about 0.1 cm but has not produced branch roots. If prepared microtome sections can be provided, this is preferable. The average length of cells in the cortex and in the vascular central portion is determined using an eyepiece graticule (see Part A). In addition to the longitudinal sections, transverse sections in a comparable region of the root are used to make cell counts in half or a quarter of a 100× magnification microscope field (see Part A).

From these measurements and cell counts, average numbers of cells produced per day (or per hour) can be computed; to gain reliable data class results should be pooled:

$$\frac{\text{rate of growth in length } (\mu\text{m day}^{-1})}{\text{average length of cells } (\mu\text{m})} \times \begin{array}{c}\text{number of}\\\text{cells in}\\\text{transverse}\\\text{section}\end{array} = \begin{array}{c}\text{number of}\\\text{cells produced}\\\text{day}^{-1}\end{array}$$

CELL DIFFERENTIATION

The progressive differentiation of the enormous number of cells produced by the growing point can be studied in cell suspensions of the first half millimetre of root tip, followed by the second half millimetre and finally a suspension of the next 1 mm of tissue. *Pisum sativum* and *Zea mays* are suitable material, but almost any other root growing from a germinated seed will do.

The cylinders of root tissue should be cut off with a sharp blade and put into 5 cm³ of a solution of 3 g CrO₃ in 100 cm³ water, vacuum infiltrated and treated as described for leaf tissue in Expt 8.5 anatomy.

The suspension of the first half millimetre of tissue will contain mainly ribbons of embryonic meristematic cells of different shapes. The second suspension will, in addition to some ribbons of embryonic cells, contain *vacuolated* meristematic cells, recently differentiated parenchyma cells and some elongated future conducting elements and fibres. In the third suspension embryonic cells will probably be absent. Differentiating cells of the cortex, and vascular and epidermal tissue will be distinguishable in this suspension.

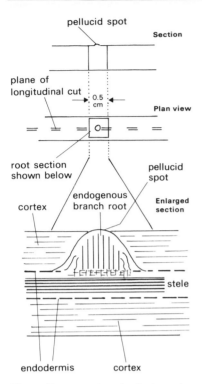

Figure B11.13c Longitudinal section of root of *Pisum sativum* with initial branch root arising endogenously. (i) Longitudinal section of portion of root; (ii) surface view of the same portion of root; (iii) longitudinal section of small length of root in region of branch root.

Root cap cells

Maize roots produce a root cap of loosely held cells of typical shape. The root cap can be removed whole with a needle or cut off with a sharp new blade, placed in a drop of 1 : 15 000 neutral red dye and gently squashed under a coverslip. The root cap cells will separate and can be observed as the dye accumulates in their vacuoles, leaving the nucleus adpressed to one side wall, unstained. Streaming and Brownian movement (see Expts 1.1, 1.14 & 3.6) can often be discerned.

Branch roots

Pisum sativum roots from seeds that have germinated in a box lined with filter paper and which have begun to produce branch roots are the recommended material. Longitudinal, hand-cut sections, about 50 μm thick, can be prepared with a sharp new razor blade. These sections should be cut through an initial branch root visible as a translucent point or a very slight swelling of the root, which may be straight or curved at that point. The sections need not be longer than 0.3 cm. It is best to cut the root near the translucent point as shown in Figure B11.13c and then shave off the desired section from one side as indicated in Figure B11.13c.

The sections are placed in a beaker with lactophenol and allowed to stand at room temperature until translucent, or they are gently heated until fumes escape, allowed to cool slightly and reheated gently until translucent (cf. Hydathodes, Part A). Often it is possible to prepare two sections from one initial branch root. The cleared sections are mounted in lactophenol and should show the features labelled in Figure B11.13c.

Part C   Technical directory

1 Recipes and formulations

acetylene

(a) If available from a cylinder, a flask should be flushed and filled with the gas, closed with a Suba-Seal and the gas then withdrawn with a syringe.

(b) If a cylinder is not available, the gas can be produced from calcium carbide and water, collected via a tube in a container, closed with a Suba-Seal (from William Freeman & Co., see Part C2) and then withdrawn with a syringe (see Fig. C1).

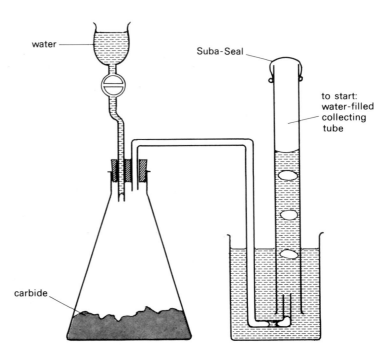

Figure C1 Longitudinal section of apparatus suitable for the production and collection of acetylene.

adjusted nutrient solution   see **nutrient solutions**.

cultures *Anabaena*   see Part A, Growing in liquid media

artificial pond water   This is prepared from 10^{-1} mol m^{-3} (100 μM) KCl, 10^{-1} mol m^{-3} (100 μM) CaCl$_2$ and 1.0 mol m^{-3} (1.0 mM) NaCl. The pH of the solution should be adjusted to 8.0.

baryta water Eighteen grammes of Analar recrystallised $Ba(OH)_2$ are dissolved in a 1000 cm³ stoppered volumetric flask filled with boiled distilled water, by allowing it to stand for 48 h. The solution may have to be filtered quickly through cotton wool. The molar concentration is determined by titration with 10^2 mol m^{-3} (0.1 M) HCl against phenolphthalein or thymol-blue indicator. Boiling the water in the volumetric flask may speed up dissolving the $Ba(OH)_2$.

A 150 mol m^{-3} (0.15 M) $Ba(OH)_2$ solution is available from Sigma Chemicals (see Part C2) and, for many experiments, the molarity required is between 3 and 5 mol m^{-3} (3 and 5 mM).

blue-green algae see **cultures** in Part A.

Brodie's solution see **manometer liquids.**

buffer solutions ACETATE–ACETIC ACID
Buffers of different pH values are prepared by mixing the two reagents in the proportions shown below.

pH	10^2 mol m^{-3} (0.1 M) sodium acetate (cm³)	10^2 mol m^{-3} (0.1 M) acetic acid (cm³)
4.0	36	164
4.6	98	102
5.0	141	59
5.4	171	29
5.6	181	19

pH values of prepared solutions must be checked and may need adjustment.

CITRATE
Buffers of different pH values are prepared by mixing solutions A and B in the proportions shown below.

Solution A is prepared by dissolving 21 g citric acid crystals in 200 cm³ of 10^3 mol m^{-3} (1.0 M) NaOH and making up to 1000 cm³ (citrate molarity 10^2 mol m^{-3} (0.1 M)).
Solution B is a 10^2 mol m^{-3} (0.1 M) NaOH solution.

pH	Solution A (cm³)	Solution B (cm³)
5.0	9.5	0.5
5.5	7.0	3.0
6.0	6.0	4.0
6.5	5.4	4.6

pH values of prepared solutions must be checked and may need adjustment.

Citrate buffer solutions of specific molarities, 1.0 mol m^{-3} (1.0 mM) and 25 mol m^{-3} (25 mM), are quoted in Experiments 11.2 and 11.5.

CITRIC ACID–SODIUM PHOSPHATE

Buffers of different pH values are prepared by mixing solutions A and B in the proportions shown below.

Solution A is 10^2 mol m^{-3} (0.1 M) citric acid·H_2O, i.e. 21 g 1000 cm^{-3}.
Solution B is 2×10^2 mol m^{-3} (0.2 M) *anhydrous* Na_2HPO_4, i.e. 28.4 g 1000 cm^{-3}.

pH	Solution A (cm^3)	Solution B (cm^3)
2.8	42.1	7.4
3.2	37.6	12.4
3.6	33.9	16.1
4.0	30.7	19.3
4.4	27.9	22.1
4.8	25.3	24.7
5.2	23.2	26.8
5.6	21.0	29.0
6.0	18.4	31.6
6.5	14.5	35.5
7.0	8.8	41.2
7.5	3.9	46.1
8.0	1.4	48.6

pH values of prepared solutions must be checked and may need adjustment.

'GOOD' BUFFER SOLUTIONS

HEPES, MOPS, MES, TES and Tris (from Sigma Chemicals, see Part C2) as formulated on the labels are weighed out according to the required molarity and the pH is adjusted with KOH or HCl.

PHOSPHATE

Buffers of different pH values are prepared by mixing solutions A and B in the proportions shown below.

Solution A is 10^2 mol m^{-3} (0.1 M) $Na_2HPO_4 \cdot 2H_2O$, i.e. 17.8 g 1000 cm^{-3}.
Solution B is 10^2 mol m^{-3} (0.1 M) $NaH_2PO_4 \cdot 2H_2O$, i.e. 15.6 g 1000 cm^{-3}.

pH	Solution A (cm^3)	Solution B (cm^3)
5.3	2.5	97.5
5.9	10.0	90.0
7.0	60.0	40.0
8.0	95.0	5.0

pH values of prepared solutions must be checked and may need adjustment.

Special phosphate buffers are quoted in Experiments 10.3 and 10.4.

Chara	This species can be kept for a few days in pond water or **artificial pond water** (see above).
Chlorella	see **cultures** in Part A.
chloroplasts	see Part A.
chromic acid	This is used for cell separation and is prepared from 3 g CrO_3 dissolved in 100 cm^3 water.
cobalt chloride papers	Three solutions, A, B and C, are required.

Solution A is 150 g $CoCl_2$ dissolved in 1000 cm^3 water.
Solution B is 1 g methylene blue dissolved in 1000 cm^3 water.
Solution C is 1 g eosin dissolved in 1000 cm^3 water.

Sheets of Whatman No. 1 filter paper are used for the three kinds of strips to be prepared. The paper must first be wetted by edge-on immersion in water for 1 min. The paper is then pressed between blotting paper and thereafter immersed for 1 min in solution A. Another sheet of paper, wetted and blotted as described above, is immersed for 1 min in ⅛th strength of solution B and a third sheet in 1/32nd strength of solution B. The last two papers are the dark and light colour standards for comparisons with the cobalt chloride paper as it changes colour during measurements.

After staining, all sheets are pressed between blotting paper and hung up to dry; the light blue standard, *when dry*, is given a second staining for 1 min in solution C, then blotted again and hung up to dry.

The strips can be arranged so that the cobalt chloride paper is between the two colour standards. For simple qualitative work the cobalt chloride paper alone is a good indicator of water vapour diffusion out of leaves. Papers should be held in Perspex or microscope slide clips.

culture media	For lower plants: see Part A, Growing in liquid media.
ground glass	This is prepared by grinding small pieces of glass in a mortar, followed by finer grinding in a glass homogeniser.
hand porometer	see **porometers.**
HEPES buffer	see **buffer solutions**, 'Good'.
hydrogen peroxide	Commercially available Analar H_2O_2 is labelled '100 vol. equivalent to about 30%'. It should be kept in the dark and dilutions prepared with distilled water to give 5, 10, 15 or 20 vol.
iodine solution	A stock solution is prepared by grinding 1 g iodine crystals together with 2 g KI in a mortar and dissolving in 300 cm^3 water. This stock solution can be diluted as required to give the desired colour.

A double strength I–KI solution is used in Experiment 2.3, but only 4 cm^3 of this are used together with a further 40 g KI to make up 1000 cm^3.

incubation medium	for isolated epidermal tissue. The medium specified below is suitable for the epidermis of *Commelina communis*; if other species are used, the concentration of potassium that will give optimum opening must be determined by a systematic experiment – a worthwhile class exercise in its own right.

For *C. communis* the recommended incubation medium is a

10 mol m^{-3} (10 mM) MES buffer adjusted to pH 6.15 with 100 mol m^{-3} (100 mM) KOH, resulting in a [K$^+$] of about 5 mol m^{-3} (5 mM). Sufficient KCl is then added to bring the total [K$^+$] to 50 mol m^{-3} (50 mM).

The tissue is incubated in the medium contained in 5-cm-diameter Petri dishes. A hypodermic syringe needle inserted through a hole in the lid of the dish is used to aerate the medium at a rate of 100 cm^3 min^{-1}. Illumination is provided either from above or below; the temperature should be kept between 20 and 25 °C. After 3 h tissue pieces are removed for microscopic measurement of pore widths of 20–30 stomata.

lactophenol   This clearing agent is made up from:

25 g phenol crystals	25 cm^3 glycerol
25 cm^3 lactic acid	25 cm^3 water

manometer liquids   Brodie's solution is prepared from 44 g Analar anhydrous NaBr dried at 100 °C and 1 cm^3 Triton-X dissolved in 1000 cm^3 water plus 0.3 g Evan's blue, if necessary filtered through sintered glass.

For resistance porometers, liquid paraffin, sp. gr. 0.86 is used.

MES buffer   see **buffer solutions**, 'Good'.

mitochondria   see Part A.

MOPS buffer   see **buffer solutions**, 'Good'.

nutrient solutions for higher plants   (Based on Long Ashton formulae.)

Stock solution	cm^3 of stock solution needed to make 1000 cm^3 of nutrient medium
Nitrate type	
50.60 g KNO$_3$	8
80.25 g Ca(NO$_3$)$_2$ anhyd.	8
46.00 g MgSO$_4$·7H$_2$O each in 1000 cm^3 water	8
52.00 g NaH$_2$PO$_4$·2H$_2$O	4
Ammonium type	
21.75 g K$_2$SO$_4$	16
55.50 g CaCl$_2$ anhyd.	8
46.00 g MgSO$_4$·7H$_2$O each in 1000 cm^3 water	8
29.75 g Na$_2$HPO$_4$·12H$_2$O	16
66.25 g (NH$_4$)$_2$SO$_4$	8
Micronutrients common to both nitrate and ammonium types	
3.30 g FeK EDTA	5
2.23 g MnSO$_4$·4H$_2$O	1
0.29 g ZnSO$_4$·7H$_2$O	1
0.25 g CuSO$_4$·5H$_2$O	1
3.10 g H$_3$BO$_3$ each in 1000 cm^3 water	1
0.12 g Na$_2$MoO$_4$·2H$_2$O	1
5.85 g NaCl	1
0.056 g CoSO$_4$·7H$_2$O	1
Adjusted nutrient solution	
Nitrogen-deficient medium	
21.75 g K$_2$SO$_4$	16
46.00 g NaH$_2$PO$_4$·2H$_2$O	4.5
46.00 g MgSO$_4$·7H$_2$O each in 1000 cm^3 water	8
80.00 g CaSO$_4$·2H$_2$O	9

When using ammonium nitrogen media with sand cultures, it is necessary to mix 0.1–0.5% $CaCO_3$ by weight in to the top half of the sand layer in order to prevent a sharp rise in pH.

Generally, to prevent changes in pH, a mixture of ammonium and nitrate type media is recommended.

Nitrogen-Deficient Medium for *water culture*
(Based on Bond's formula.)

$$\left.\begin{array}{l}
0.87 \text{ g } K_2SO_4 \\
0.50 \text{ g } MgSO_4 \cdot 7H_2O \\
0.25 \text{ g } Ca_3(PO_4)_2 \\
0.50 \text{ g } CaSO_4 \cdot 2H_2O \\
0.25 \text{ g } Fe_3(PO_4)_2 \cdot 8H_2O
\end{array}\right\} \text{ in } 1000 \text{ cm}^3 \text{ water}$$

The calcium salts are sparingly soluble.

plasmodesmata   see page 50.

pond water   see **artificial pond water**.

porometers   Figure C3 shows a hand porometer (Meidner & Mansfield 1968). After clamping it on to a leaf without the pipette bulb and ground-glass connection, the bulb is compressed between thumb and forefinger and the ground-glass connection fitted to the clamp. The time required for the bulb to re-inflate when the pressure is released is a measure of stomatal conductance; it can be timed with a watch or by counting.

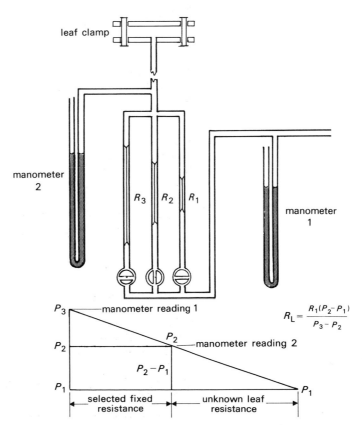

$$R_L = \frac{R_1(P_2 - P_1)}{P_3 - P_2}$$

Figure C2 The resistance porometer in longitudinal section and the pressure relations established during the operation of the instrument. (See Meidner, H. and T. A. Mansfield 1968. *Physiology of stomata*. London: McGraw-Hill.)

Figure C2 shows a resistance porometer. The theory underlying its use is also indicated in Figure C2.

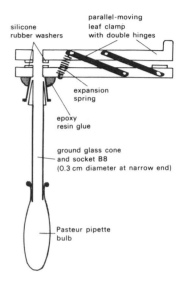

silicone
rubber washers

parallel-moving
leaf clamp
with double hinges

expansion
spring

epoxy
resin glue

ground glass cone
and socket B8
(0.3 cm diameter at narrow end)

Pasteur pipette
bulb

Figure C3 A hand porometer in longitudinal section. The clamp action can also be achieved by a compression spring on a centrally placed boss fixed to the upper jaw and passing loosely through the lower. (See Meidner, H. and T. A. Mansfield 1968. *Physiology of stomata*. London: McGraw-Hill).

protoplasts	see Part A.
resistance porometer	see **porometers**.
soluble starch	Different strengths of soluble starch sols are needed for different experiments, but sols of between 0.5 and 2 g of starch 100 cm⁻³ are usual. The starch must first be made into a paste with a few drops of water and the paste then added to 75 cm³ of boiling water under stirring and finally made up to 100 cm³. In Experiment 2.1 the starch paste is added to a boiling agar sol.
solute potential of sucrose solutions	Although a non-electrolyte and not ionising, molar sucrose solutions have irregular osmotic properties. Figure C4 allows for the appropriate values to be read off for concentrations between 50 mol m⁻³ and 1 × 10³ mol m⁻³ (50 mM and 1.0 M). Values are valid for solutions at 20 °C.
stains	The following concentrations are usually used:

eosin	0.1 g 500 cm⁻³
methylene blue	0.1 g 500 cm⁻³
neutral red	0.1 g 1000 cm⁻³ is usual, but double and half strengths can be employed
ruthenium red	0.1 g 500 cm⁻³, specific for middle lamellae
saffranin, aqueous	0.25 g 100 cm⁻³

starch	see **soluble starch**.
sucrose solutions	see **solute potential**.
suspensions	for *Anabaena*, *Chlorella*, *Nostoc*, *Saccharomyces* and *Scenedesmus*, see **cultures** in Part A. For chloroplasts, mitochondria and protoplasts, see Part A.
TES buffer	see **buffer solutions**, 'Good'.

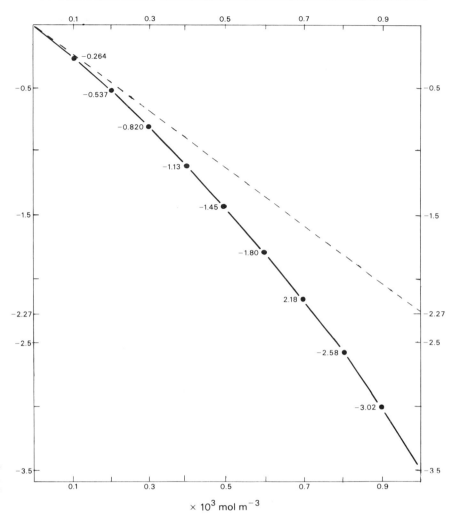

Figure C4 The relation between the molar concentration of sucrose solutions and their ψ_{solute} (10^3 mol m^{-3} = 1.0 M).

| **Tris buffer** | see **buffer solutions**, 'Good'. |
| **yeasts** | see **cultures**: *Saccharomyces* in Part A. |

REFERENCE

Meidner, H. and T. A. Mansfield 1968. *Physiology of stomata*. London: McGraw-Hill.

3 Names and addresses of contributors

Miss Gail Alexander, Department of Biological Sciences, University of Dundee, Dundee DD1 4HN, UK

Professor D. A. Baker, Department of Biological Sciences, Wye College, Ashford TN25 5AH, UK

Professor J. Barber, Department of Pure and Applied Biology, Imperial College of Science and Technology, Prince Consort Rd, London SW7 2BB, UK

Dr. A. M. Barclay, Department of Botany, University of St Andrews, St Andrews, KY16 9AL, UK

Dr D. J. F. Bowling, Department of Botany, University of Aberdeen, St Machar's Drive, Aberdeen AB9 2UD, UK

Dr H. A. Collin, Department of Botany, University of Liverpool, POB 147, Liverpool L69 38X, UK

Dr J. Collins, Department of Botany, University of Liverpool, POB 147, Liverpool L69 38X, UK

Professor R. M. M. Crawford, Department of Botany, University of St Andrews, St Andrews KY16 9AL, UK

Dr W. J. Davies, Department of Biological Sciences, University of Lancaster, Lancaster LA1 4YQ, UK

Dr M. Donkin, Department of Biological Sciences, Plymouth Polytechnic, Plymouth PL4 8AA, UK

Mr P. St J. Edwards, Department of Biology, University of Salford, Salford M5 4W9, UK

Dr J. F. Farrar, Department of Plant Biology, University College Bangor, Bangor LL57 2UW, UK

Dr P. J. Fitzsimons, Department of Biological Sciences, University of Dundee, Dundee DD1 4HN, UK

Dr A. Goldsworthy, Department of Pure and Applied Biology, Imperial College of Science and Technology, Prince Consort Rd, London SW7 2BB, UK

Professor M. A. Hall, Department of Botany and Microbiology, University College Aberystwyth, Aberystwyth SY23 3DA, UK

Dr J. Hannay, Department of Pure and Applied Biology, Imperial College of Science and Technology, Prince Consort Rd, London SW7 2BB, UK

Dr K. Hardwick, Department of Botany, University of Liverpool, POB 147, Liverpool L69 38X, UK

Dr R. Hunt, Unit of Comparative Plant Ecology (NERC), Department of Botany, University of Sheffield, Sheffield S10 2TN, UK

Dr D. Idle, Department of Plant Biology, University of Birmingham, Birmingham B15 2TT, UK

Professor D. H. Jennings, Department of Botany, University of Liverpool, POB 147, Liverpool L69 38X, UK

Dr R. O. Mackender, Department of Botany, Queen's University, Belfast BT7 1NN, UK

Dr C. C. McCready, Department of Botany, University of Oxford, South Parks Rd, Oxford OX1 3RA, UK

Dr E. S. Martin, Department of Biological Sciences, Plymouth Polytechnic, Plymouth PL4 8AA, UK

Professor T. A. Mansfield, Department of Biological Sciences, University of Lancaster, Lancaster LA1 4YQ, UK

Professor H. Meidner, Department of Biological Sciences, University of Stirling, Stirling FK9 4LA, UK

Dr P. W. Mueller, Department of Pure and Applied Biology, Imperial College of Science and Technology, Prince Consort Rd, London SW7 2BB, UK

Dr J. M. Palmer, Department of Pure and Applied Biology, Imperial College of Science and Technology, Prince Consort Rd, London SW7 2BB, UK

Professor J. S. Pate, Department of Botany, University of Western Australia, Nedlands, Western Australia, 6009

Dr R. Phillips, Department of Biological Sciences, University of Stirling, Stirling FK9 4LA, UK

Dr D. N. Price, Department of Biological Sciences, Plymouth Polytechnic, Plymouth PL4 8AA, UK

Dr R. Sexton, Department of Biological Sciences, University of Stirling, Stirling FK9 4LA, UK

Professor E. W. Simon, Department of Botany, Queen's University, Belfast BT7 1NN, UK

Dr A. R. Smith, Department of Botany and Microbiology, University College Aberystwyth, Aberystwyth SY23 3DA, UK

Professor W. D. P. Stewart, Department of Biological Sciences, University of Dundee, Dundee DD1 4HN, UK

Dr J. D. B. Weyers, Department of Biological Sciences, University of Dundee, Dundee DD1 4HN, UK

Dr C. Willmer, Department of Biological Sciences, University of Stirling, Stirling FK9 4LA, UK

Dr J. M. Wilson, Department of Plant Biology, University College Bangor, Bangor LL57 2UW, UK

Dr M. J. Wren, Department of Plant Sciences, University of Leeds, Leeds LS2 9JT, UK

Index